Pitman Research Notes in Mathematics Series

Submission of proposals for consideration

Suggestions for publication, in the form of outlines and representative samples, are invited by the Editorial Board for assessment. Intending authors should approach one of the main editors or another member of the Editorial Board, citing the relevant AMS subject classifications. Alternatively, outlines may be sent directly to the publisher's offices. Refereeing is by members of the board and other mathematical authorities in the topic concerned, throughout the world.

Preparation of accepted manuscripts

On acceptance of a proposal, the publisher will supply full instructions for the preparation of manuscripts in a form suitable for direct photo-lithographic reproduction. Specially printed grid sheets are provided and a contribution is offered by the publisher towards the cost of typing. Word processor output, subject to the publisher's approval, is also acceptable.

Illustrations should be prepared by the authors, ready for direct reproduction without further improvement. The use of hand-drawn symbols should be avoided wherever possible, in order to maintain maximum clarity of the text.

The publisher will be pleased to give any guidance necessary during the preparation of a typescript, and will be happy to answer any queries.

Important note

In order to avoid later retyping, intending authors are strongly urged not to begin final preparation of a typescript before receiving the publisher's guidelines and special paper. In this way it is hoped to preserve the uniform appearance of the series.

Longman Scientific & Technical
Longman House
Burnt Mill
Harlow, Essex, UK
(tel (0279) 26721)

W9-AFD-400

Titles in this series

Completely bounded
maps and dilations

Vern I Paulsen

University of Houston, University Park

Completely bounded maps and dilations

Longman
Scientific &
Technical

Copublished in the United States with
John Wiley & Sons, Inc., New York

Longman Scientific & Technical
Longman Group UK Limited
Longman House, Burnt Mill, Harlow
Essex CM20 2JE, England
and Associated Companies throughout the world.

Copublished in the United States with
John Wiley & Sons, Inc., 605 Third Avenue, New York, NY 10158

First published 1986

AMS Subject Classifications: (main) 46L05, 47A20, 47A25
                            (subsidiary) 47C15

ISSN 0269-3674

British Library Cataloguing in Publication Data

Paulsen, Vern
    Completely bounded maps and dilations.—
    (Pitman research notes in mathematics,
    ISSN 0269-3674; 146)
    1. Analytic functions
    I. Title
    515'.223   QA331
    ISBN 0-582-98896-9

Library of Congress Cataloging-in-Publication Data
Paulsen, Vern, 1951–
    Completely bounded maps and dilations.
    (Pitman research notes in mathematics series,
ISSN 0269-3674; 146)
    Bibliography: p.
    Includes index.
    1. Dilation theory (Operator theory)   2. Mappings
(Mathematics)   3. C*-algebras. I. Title. II. Series:
Pitman research notes in mathematics; 146.
QA329.P38 1986   515.7'24   86-15697
ISBN 0-470-20369-2 (USA only)

Printed and bound in Great Britain by
Biddles Ltd., Guildford and Kings Lynn

Preface

A standard technique for studying operators on a Hilbert space is to realize an operator as part of a simpler operator whose domain is a larger Hilbert space. This essentially geometric technique is referred to as dilation. One of the better known theorems of this type is due to Sz.-Nagy and asserts that any contraction operator can be dilated to a unitary operator in such a fashion that the powers of the unitary also simultaneously dilate the powers of the contraction. Such dilations were systematically studied in Sz.-Nagy and Foias', "Harmonic Analysis of Operators on Hilbert Space".

Completely positive maps were introduced by Stinespring as an algebraic setting for studying the existence of dilations. The connections between completely positive maps and dilation theory were broadened further by Arveson, who developed a deep structure theory for these maps, including an operator valued Hahn-Banach extension theory.

Completely positive maps also play a central role in the theory of tensor products of C^*-algebras and in non-commutative harmonic analysis. Characterizations of nuclear C^*-algebras and injective factors are given in terms of these maps, and in harmonic analysis they arise in the guise of positive definite operator valued functions on groups. In spite of the broad range of applications of this class of maps, there is no concise treatment of them.

Recent work has shown that much of the theory of completely positive maps can be quite easily extended to a considerably broader class of maps, the completely bounded maps. The completely bounded maps play a role in the study of skewed dilations that is analogous to the role that completely positive maps play in dilation theory. A skewed dilation is a dilation except that there are two embeddings of the original Hilbert space in the larger space. A well-known example of this type of construction is due to Rota, who proved that if the spectrum of an operator is contained in the open unit disc, then the operator can be dilated in a skewed fashion to a unitary operator.

Rota used the construction of this skewed dilation to prove that every operator whose spectrum is contained in the open unit disc is similar to a contraction operator. The technique of deducing the existence of a similarity by first constructing a skewed dilation is often used in operator theory. The correspondence between skewed dilations and completely bounded maps makes these maps the unifying concept behind many results on the existence of similarities. Characterizations of the operators that are similar to contractions, the representations of C^*-algebras that are similar to *-representations, and the derivations of C^*-algebras that are inner, have all been given necessary and sufficient conditions in terms of completely bounded maps.

The purpose of this set of notes is to present a unified introduction to the theories of the completely positive and completely bounded maps. We include as many of the applications of these theories to dilation theory and similarity theory as is possible in a relatively short space. Our particular emphasis is on applications to operator theory.

These notes are intended to be accessible to a student who has had a basic course in operator theory including an introduction to Banach algebras and C^*-algebras.

Contents

Chapter 1 is a brief introduction to the concept of completely positive and completely bounded maps. In Chapter 2 we study positive maps and their relationships with unital, contractive maps. We use these ideas to give an elementary proof of von Neumann's inequality and obtain the Russo-Dye theorem as a corollary.

Chapter 3 contains some introductory material on completely positive maps and obtains the Berger-Kato-Stampfli result on operators with numerical radius 1 as an application of these ideas.

Chapter 4 is devoted to the dilation theorems of Stinespring, Sz.-Nagy, Naimark, Berger, and Berger-Foias-Lebow. The correspondence between positive definite and completely positive definite functions on a group is studied as well as bimodule maps.

The duality between completely positive maps into an $n \times n$ matrix algebra and linear functionals on an associated space is explored in Chapter 5. This yields a proof of Arveson's extension theorem in the finite dimensional case and allows us to analyze the difference between unital, positive maps and unital, contractive maps.

Chapter 6 contains the full version of Arveson's extension theorem as well as his analysis of dilation questions in terms of unital, completely contractive maps. Parrott's example of three commuting contractions which fail to have a commuting unitary dilation is studied.

In Chapter 7 a skewed dilation theory for completely bounded maps is developed and it is used to generalize many parts of the theory of completely positive maps to this larger class. In particular, Wittstock's decomposition theorem and the Hahn-Banach extension theorem for completely bounded maps are proved.

Chapter 8 studies completely bounded homomorphisms and their applications to similarity questions.

Chapter 9 presents the model theory for operators which have a finitely connected region as a K-spectral set that was developed jointly with Douglas.

Finally, Chapter 10 contains an introduction to the theory of tensor products and gives some applications of these ideas to the study of joint K-spectral sets for several commuting operators.

Acknowledgements

These notes grow out of a series of lectures given at the University of Houston and SUNY at Stony Brook in 1982 and 1983.

The works and ideas of many people have influenced these notes. The main influences have been: W. B. Arveson, M. D. Choi, R. G. Douglas, E. G. Effros, U. Haagerup, and G. Wittstock.

Additional thanks go to J. McCarthy, W. Paschke, R. Smith, C. Y. Suen, and J. Ward who suffered through earlier versions of these notes and have made many technical and expository contributions.

Finally, thanks go to my typist Cindy Pavlica, whose patience and diligence at times seemed limitless, and to Sue, who gave up her dining room table for this project.

To John

1 Introduction

It is assumed throughout these notes that the reader is familiar with operator theory and the basic properties of C^*-algebras (see for example [35] and [5, Chapter 1]). We concentrate primarily on giving a self-contained exposition of the properties of completely positive and completely bounded maps between C^*-algebras and the applications of these maps to similarities and dilation theory. In particular, we assume that the reader is familiar with the material necessary for the Gelfand-Naimark-Segal theorem, which states that every C^*-algebra has a one-to-one, *- preserving, norm-preserving representation as a norm closed, *-closed algebra of operators on a Hilbert space.

As well as having a norm a C^*-algebra also has an order structure, induced by the cone of positive elements. Recall that an element of a C^*-algebra is positive if and only if it is self-adjoint and its spectrum is contained in the non-negative reals, or equivalently, if it is of the form a^*a for some element a . Since the property of being positive is preserved by *-isomorphisms, if a C^*-algebra is represented as an algebra of operators on a Hilbert space then the positive elements of the C^*-algebra coincide with the positive operators that are contained in the representation of the algebra. An operator A is positive provided that the inner product, $<Ax,x>$ is non-negative for every vector x in the space. We shall also write $a \geq 0$ to denote that a is positive.

The positive elements in a C^*-algebra A are a norm closed, convex cone in the C^*-algebra, denoted by A^+ . If h is a self-adjoint element, then it is easy to see via the functional calculus, that h is the difference of two positive elements. Indeed, if we let

$$f^+(x) = \begin{cases} x, & x \geq 0 \\ 0, & x < 0 \end{cases} \quad , \quad f^-(x) = \begin{cases} 0, & x \geq 0 \\ -x, & x < 0 \end{cases} \quad ,$$

then $h = f^+(h) - f^-(h)$, with $f^+(h)$ and $f^-(h)$ both positive. In particular, we see that the real linear span of the positive elements is the set of self-adjoint elements, which is also norm closed.

Using the Cartesian decomposition of an arbitrary element a of A , $a = h + ik$ with $h = h^*$, $k = k^*$, we see that

$$a = (p_1-p_2) + i(p_3-p_4) \ ,$$

with p_i positive, $i = 1, 2, 3, 4$.

In addition to having its own norm and order structure, a C^*-algebra also yields a whole sequence of norms and order structures on a set of C^*-algebras naturally associated with the original algebra. To see this, let A be our C^*-algebra, let M_n denote the n × n complex matrices, and let $M_n(A)$ denote the set of n × n matrices with entries from A . We'll denote a typical element of $M_n(A)$ by $(a_{i,j})$.

There is a natural way to make $M_n(A)$ into a *-algebra. Namely for $(a_{i,j})$ and $(b_{i,j})$ in $M_n(A)$, set

$$(a_{i,j}) \cdot (b_{i,j}) = (\Sigma^n_{k=1} a_{i,k} b_{k,j}) \ , \quad \text{and}$$
$$(a_{i,j})^* = (a^*_{j,i}) \ .$$

What is not so obvious is that there is a unique way to introduce a norm such that $M_n(A)$ becomes a C^*-algebra.

One way that $M_n(A)$ can be viewed as a C^*-algebra is to first choose a one-to-one *-representation of A on some Hilbert space H , and then

2

let $M_n(A)$ act on the direct sum of n copies of H in the obvious way. It is easy to verify that this defines a one-to-one representation of $M_n(A)$ for which the above multiplication and *-operation become operator composition and operator adjoint. It is straightforward to verify that the image of $M_n(A)$ under this representation is closed and hence a C^*-algebra.

Thus, we have a way to turn $M_n(A)$ into a C^*-algebra. But since the norm is unique on a C^*-algebra, we see that the norm on $M_n(A)$ defined in this fashion is independent of the particular representation of A that we chose. Since positive elements remain positive under *-isomorphisms, we see that the positive elements of $M_n(A)$ are also uniquely determined.

For some examples of this construction, first consider M_k. We can regard this as a C^*-algebra by identifying M_k with the linear transformations on k-dimensional (complex) Hilbert space, $\mathbb{C}^k$. There is a natural way to identify $M_n(M_k)$ with M_{nk}, namely, forget the additional parentheses. It is easy to see that, with this identification, the multiplication and *-operation on $M_n(M_k)$ become the usual multiplication and *-operation on M_{nk}, that is, the identification defines a *-isomorphism. Hence, the unique norm on $M_n(M_k)$ is just the norm on M_{nk}. An element of $M_n(M_k)$ will be positive if and only if the corresponding matrix in M_{nk} is positive.

For a second example, let X be a compact Hausdorff space and let $C(X)$ denote the continuous complex-valued functions on X. Setting
$$f^*(x) = \overline{f(x)} \, ,$$

$$\| f \| = \sup \{|f(x)| : x \in X\} \, ,$$

and defining the algebra operations pointwise makes $C(X)$ into a C^*-algebra. An element $F = (f_{i,j})$ of $M_n(C(X))$ can be thought of as a

continuous M_n-valued function. Note that the addition, multiplication,
and *-operation in $M_n(C(X))$ are just the pointwise addition, pointwise
multiplication, and pointwise conjugate-transpose operations of these
matrix-valued functions. If we set

$$F = \sup \{\|F(x)\| : x \in X\} ,$$

where by $\|F(x)\|$ we mean the norm in M_n , then it is easily seen that this
defines a C^*-norm on $M_n(C(X))$, and thus is the unique norm in which
$M_n(C(X))$ is a C^*-algebra. Note that the positive elements of $M_n(C(X))$
are those F for which $F(x)$ is a positive matrix for all x .

There is an alternative approach to the above constructions via tensor
products. The astute reader has perhaps realized that the algebra $M_n(A)$
that we've defined is readily identified with the tensor product algebra
$M_n \otimes A$. If $\{E_{i,j}\}_{i,j=1}^n$ denotes the canonical basis for M_n , then an
element $(a_{i,j})$ in $M_n(A)$ corresponds to $\sum_{i,j=1}^n a_{i,j} \otimes E_{i,j}$ in $M_n \otimes A$.
Recall that one makes the tensor product of two algebras into an algebra
by defining $(a_1 \otimes b_1) \cdot (a_2 \otimes b_2) = (a_1 a_2) \otimes (b_1 b_2)$ and then extending
linearly. We leave it to the reader to verify that with the above
identification of $M_n(A)$ and $M_n \otimes A$, the multiplication defined on $M_n(A)$
becomes the tensor product multiplication on $M_n \otimes A$.

We shall on occasion return to this tensor product notation to
simplify concepts.

Now, given two C^*-algebras A and B and a map $\phi: A \to B$, we also
obtain maps $\phi_n: M_n(A) \to M_n(B)$ via the formula

$$\phi_n((a_{i,j})) = (\phi(a_{i,j})) .$$

In general the adverb completely means that all of the maps $\{\phi_n\}$ enjoy
some property.

4

Thus, the map ϕ is called <u>positive</u> if it maps positive elements of A to positive elements of B and <u>completely positive</u> if every ϕ_n is a positive map.

In a similar fashion, if ϕ is a bounded map, then each ϕ_n will be bounded, and when $\sup_n \|\phi_n\|$ is finite, we call ϕ a <u>completely bounded</u> map.

One's initial reaction is perhaps that C^*-algebras are sufficiently nice that every positive map is completely positive and every bounded map is completely bounded. Indeed, one might expect that $\|\phi\| = \|\phi_n\|$ for all n . For these reasons, we begin with an example of a fairly nice map where one encounters this difference.

Let $\{E_{i,j}\}^2_{i,j=1}$ denote the system of matrix units for M_2 , that is, $E_{i,j}$ is 1 in the (i,j)-th entry and 0 elsewhere and let $\phi \colon M_2 \to M_2$ be the transpose map. It is easy to verify that the transpose of a positive matrix is positive and that the norm of the transpose of a matrix is the same as the norm of the matrix, so ϕ is positive and $\|\phi\| = 1$. Now let's consider $\phi_2 \colon M_2(M_2) \to M_2(M_2)$.

Note that the matrix of matrix units,

$$\begin{bmatrix} E_{11} & E_{12} \\ E_{21} & E_{22} \end{bmatrix} = \begin{bmatrix} 1 & 0 & 0 & 1 \\ 0 & 0 & 0 & 0 \\ 0 & 0 & 0 & 0 \\ 1 & 0 & 0 & 1 \end{bmatrix}$$

is positive, but that

$$\phi_2 \left(\begin{bmatrix} E_{11} & E_{12} \\ E_{21} & E_{22} \end{bmatrix} \right) = \begin{bmatrix} \phi(E_{11}) & \phi(E_{12}) \\ \phi(E_{21}) & \phi(E_{22}) \end{bmatrix} = \begin{bmatrix} 1 & 0 & 0 & 0 \\ 0 & 0 & 1 & 0 \\ 0 & 1 & 0 & 0 \\ 0 & 0 & 0 & 1 \end{bmatrix}$$

5

is not positive. Thus, ϕ is positive but not completely positive. In a similar fashion, we have that

$$\left\| \begin{bmatrix} E_{11} & E_{21} \\ E_{12} & E_{22} \end{bmatrix} \right\| = 1$$

while the norm of its image under ϕ_2 has norm 2. Thus, $\|\phi_2\| \geq 2$, so $\|\phi_2\| \neq \|\phi_1\|$. It turns out that ϕ is completely bounded, in fact, $\sup_n \|\phi_n\| = 2$, as we shall see later.

To obtain an example of a map that's not completely bounded, we need to repeat the above example but on an infinite dimensional space. So let H be a separable, infinite dimensional Hilbert space with a countable, orthonormal basis, $\{e_n\}_{n=1}^{\infty}$. Every bounded, linear operator T on H can be thought of as an infinite matrix whose (i,j)-th entry is the inner product $\langle Te_j, e_i \rangle$. One then defines a map ϕ from the C^*-algebra of bounded linear operators on H, $L(H)$, to $L(H)$ by the transpose. Again ϕ will be positive and an isometry, but $\|\phi_n\| \geq n$. To see this last claim, let $\{E_{i,j}\}_{i,j=1}^{\infty}$ be matrix units on H, and for fixed n, let $A = (E_{j,i})$, that is, A is the element of $M_n(L(H))$ whose (i,j)-th entry is $E_{j,i}$. We leave it to the reader to verify that $\|A\| = 1$ (in fact, A is a partial unitary), but $\|\phi_n(A)\| = n$ (Exercise 1.4).

Before turning our attention to the completely positive or completely bounded maps, we begin with some results on positive maps which we shall need repeatedly. These results also serve to illustrate how many simplifications arise when one passes to this smaller class of maps.

6

EXERCISES

1.1. Let A and B be unital C^*-algebras, and let $\pi: A \to B$ be a *-homomorphism with $\pi(1) = 1$. Show that π is completely positive and completely bounded and that $\|\pi\| = \|\pi_n\| = \|\pi\|_{cb} = 1$.

1.2. Let A, B, and C be C^*-algebras, and let $\phi: A \to B$ and $\psi: B \to C$ be (completely) positive maps. Show that $\psi \circ \phi$ is (completely) positive.

1.3. Let $\{E_{i,j}\}_{i,j=1}^n$ be matrix units for M_n, let $A = (E_{j,i})_{i,j=1}^n$ and let $B = (E_{i,j})_{i,j=1}^n$ be in $M_n(M_n)$. Show that A is unitary and that $\frac{1}{n} B$ is a rank one projection.

1.4. Let $\{E_{i,j}\}_{i,j=1}^\infty$ be a system of matrix units for $L(H)$, let $A = (E_{j,i})_{i,j=1}^n$, and let $B = (E_{i,j})_{i,j=1}^n$ be in $M_n(L(H))$. Show that A is a partial unitary, and that $\frac{1}{n} B$ is a projection. Show that $\phi_n(A) = B$ and $\|\phi_n(A)\| = n$.

2 Positive maps

If S is a subset of a C^*-algebra A, then we set

$$S^* = \{a: a^* \in S\} \,,$$

and we call S self-adjoint when $S = S^*$. If A has a unit 1 and S is a self-adjoint subspace of A containing 1, then we call S an operator system. If S is an operator system and h is a self-adjoint element of S, then even though $f^+(h)$ and $f^-(h)$ need not belong to S, we can still write h as the difference of two positive elements in S. Indeed,

$$h = \frac{1}{2} (\|h\| \cdot 1 + h) - \frac{1}{2} (\|h\| \cdot 1 - h) \,.$$

If S is an operator system, B is a C^*-algebra and $\phi: S \rightarrow B$ is a linear map, then ϕ is called positive provided that it maps positive elements of S to positive elements of B. In this chapter, we develop some of the properties of positive maps. In particular, we shall be concerned with how the assumption of positivity is related to the norm of the map, and conversely, when assumptions about the norm of a map guarantee that it is positive. We give a fairly elementary proof of von Neumann's inequality (Corollary 2.7), which only uses these observations about positive maps and an elementary result from complex analysis due to Fejer and Riesz.

If ϕ is a positive, linear functional on an operator system S, then it is easy to show that $\|\phi\| = \phi(1)$ (Exercise 2.3). When the range is a C^*-algebra the situation is quite different.

8

Proposition 2.1. Let S be an operator system and let B be a unital C^*-algebra. If $\phi: S \to B$ is a positive map, then ϕ is bounded and $\|\phi\| \leq 2\|\phi(1)\|$.

Proof. First note that if p is positive, then $0 \leq p \leq \|p\| \cdot 1$ and so, $0 \leq \phi(p) \leq \|p\| \cdot \phi(1)$ from which it follows that $\|\phi(p)\| \leq \|p\| \cdot \|\phi(1)\|$ when $p \geq 0$.

Next note that if p_1 and p_2 are positive, then $\|p_1 - p_2\| \leq \max \{\|p_1\|, \|p_2\|\}$. If h is self-adjoint in S , then using the above decomposition of h , we have

$$\phi(h) = \frac{1}{2}\phi(\|h\| \cdot 1 + h) - \frac{1}{2}\phi(\|h\| \cdot 1 - h) ,$$

which expresses $\phi(h)$ as a difference of two positive elements of B . Thus,

$$\|\phi(h)\| \leq \frac{1}{2} \max \{\|\phi(\|h\| \cdot 1 + h)\|, \|\phi(\|h\| \cdot 1 - h)\|\} \leq \|h\| \cdot \|\phi(1)\| .$$

Finally, if a is an arbitrary element of S , then $a = h + ik$ with $\|h\|, \|k\| \leq \|a\|$, $h = h^*$, $k = k^*$, and so,

$$\|\phi(a)\| \leq \|\phi(h)\| + \|\phi(k)\| \leq 2 \|a\| \cdot \|\phi(1)\| . \qquad \square$$

Let us reproduce an example of Arveson which shows that 2 is the best constant in the above Proposition.

Example 2.2. Let Π denote the unit circle in the complex plane, $C(\Pi)$ the continuous functions on Π , z the coordinate function, and $S \subseteq C(\Pi)$ the subspace spanned by 1 , z , and $\bar{z}$.

We define $\phi: S \to M_2$ by

$$\phi(a + bz + c\bar{z}) = \begin{bmatrix} a & 2b \\ 2c & a \end{bmatrix} .$$

We leave it to the reader to verify that an element, $a1 + bz + c\bar{z}$ of S is positive if and only if $c = \bar{b}$ and $a \geq 2|b|$. It is fairly standard that a self-adjoint element of M_2 is positive if and only if its diagonal entries and its determinant are non-negative real numbers. Combining these two facts it is clear that ϕ is a positive map. However,

$$2\|\phi(1)\| = 2 = \|\phi(z)\| \leq \|\phi\| ,$$

so that $\|\phi\| = 2\|\phi(1)\|$.

The existence of unital, positive maps which are not contractive can be roughly attributed to two factors. One is the non-commutativity of the range, the other is the "lack" of sufficiently many positive elements in the domain. This first principle is illustrated in the exercises and we concentrate here on properties of the domain which ensure that unital, positive maps are contractive.

Lemma 2.3. Let A be a C^*-algebra with unit and let p_i , $i = 1, \ldots, n$, be positive elements of A such that

$$\Sigma^n_{i=1} p_i \leq 1 .$$

If λ_i, $i = 1, \ldots, n$, are scalars with $|\lambda_i| \leq 1$, then

$$\|\Sigma^n_{i=1} \lambda_i p_i\| \leq 1 .$$

Proof. Note that

$$
\begin{bmatrix} \Sigma^n_{i=1}\lambda_i p_i & 0 & \cdot & \cdot & 0 \\ 0 & 0 & & & \cdot \\ \cdot & & \cdot & & \cdot \\ \cdot & & & \cdot & \cdot \\ 0 & \cdot & \cdot & \cdot & \cdot & 0 \end{bmatrix}
=
\begin{bmatrix} p_1^{\frac{1}{2}} & \cdots & p_n^{\frac{1}{2}} \\ 0 & \cdots & 0 \\ \cdot & & \cdot \\ \cdot & & \cdot \\ 0 & \cdots & 0 \end{bmatrix}
\cdot
\begin{bmatrix} \lambda_1 0 & \cdots & 0 \\ 0 & \cdot & & \cdot \\ \cdot & \cdot & & \cdot \\ \cdot & & \cdot & 0 \\ 0 & \cdots & 0 \lambda_n \end{bmatrix}
\cdot
\begin{bmatrix} p_1^{\frac{1}{2}} & 0. & .0 \\ \cdot & \cdot & \cdot \\ \cdot & \cdot & \cdot \\ \cdot & \cdot & \cdot \\ p_n^{\frac{1}{2}} & 0. & .0 \end{bmatrix}
.
$$

The norm of the matrix on the left is $\|\Sigma_{i=1}^{n} \lambda_i p_i\|$, while each of the three matrices on the right can be easily seen to have norm less than 1 , by using the fact that $\|a^* a\| = \|aa^*\| = \|a\|^2$. □

Theorem 2.4. Let $\mathcal{B}$ be a C^*-algebra with unit, let X be a compact Hausdorff space, $C(X)$ the continuous functions on X , and let $\phi: C(X) \to \mathcal{B}$ be a positive map. Then $\|\phi\| = \|\phi(1)\|$.

Proof. Clearly, we may assume that $\phi(1) \leq 1$. Let $f \in C(X)$, $\|f\| \leq 1$, and let $\varepsilon > 0$ be given. First, we note that be a standard partition of unity argument, f may be approximated to within ε by a sum of the form given in Lemma 2.3. To see this, first choose a finite open covering $\{U_i\}_{i=1}^{n}$ of X such that $|f(x) - f(x_i)| < \varepsilon$ for x in U_i , and let $\{p_i\}$ be a partition of unity subordinate to the covering. Set $\lambda_i = f(x_i)$ and note that if $p_i(x) \neq 0$ for some i , then $|f(x) - \lambda_i| < \varepsilon$. Hence, for any x ,

$$|f(x) - \Sigma\lambda_i p_i(x)| = |\Sigma(f(x) - \lambda_i)p_i(x)| < \Sigma|f(x) - \lambda_i|p_i(x) < \Sigma\varepsilon \cdot p_i(x) = \varepsilon.$$

Finally, by Lemma 2.3, $\|\Sigma \lambda_i\phi(p_i)\| \leq 1$, so that $\|\phi(f)\| \leq \|\phi(f - \Sigma \lambda_i p_i)\| + \|\Sigma \lambda_i\phi(p_i)\| < 1 + \varepsilon \cdot \|\phi\|$, and since ε was arbitrary, $\|\phi\| \leq 1$. □

As an application of Theorem 2.4, we now prove an inequality due to von Neumann, which will be central to many later results.

Lemma 2.5 (Fejer-Riesz). Let $\tau(e^{i\theta}) = \Sigma_{n=-N}^{+N} a_n e^{in\theta}$ be a positive function on the unit circle Π . Then there is a polynomial p such that

$$\tau(e^{i\theta}) = |p(e^{i\theta})|^2 .$$

<u>Proof</u>. First note that since τ is real-valued, $a_{-n} = \bar{a}_n$ and a_o is real. Set $g(z) = \Sigma_{n=-N}^{+N} a_n z^{n+N}$, so that g is a polynomial of degree $2N$ with $g(0) \neq 0$. Notice that the anti-symmetry of the coefficients of g implies

$$\overline{g(1/\bar{z})} = z^{-2N} g(z) .$$

This means that the $2N$ zeroes of g may be written as $z_1, \ldots, z_N$, $1/\bar{z}_1, \ldots, 1/\bar{z}_N$.

We set $q(z) = (z - z_1)\cdots(z - z_N)$, $h(z) = (z - 1/\bar{z}_1)\cdots(z - 1/\bar{z}_N)$, and have that

$$g(z) = a_N q(z) h(z) ,$$

with

$$\overline{h(z)} = \frac{(-1)^N z^{-N} q(1/\bar{z})}{z_1 \cdots z_N} .$$

Thus, $\tau(e^{i\theta}) = e^{-iN\theta} g(e^{i\theta}) = |g(e^{i\theta})| = |a_N| \cdot |q(e^{i\theta})| \cdot |\overline{h(e^{i\theta})}| =$

$\left|\dfrac{a_N}{z_1 \cdots z_N}\right| \cdot |q(e^{i\theta})|^2$, so that $\tau(e^{i\theta}) = |p(e^{i\theta})|^2$, where

$$p(z) = \left|\frac{a_N}{z_1 \cdots z_N}\right|^{\frac{1}{2}} q(z) . \qquad \square$$

Writing $p(z) = \alpha_o + \ldots + \alpha_N z^N$, we see that $\tau(e^{i\theta}) = \Sigma_{\ell,k=0}^N \alpha_\ell \bar{\alpha}_k e^{i(\ell-k)\theta}$, so that the coefficients of every positive trigonometric polynomial have this special form.

12

Theorem 2.6. Let T be an operator on a Hilbert space with $\|T\| \leq 1$ and let $S \subseteq C(\Pi)$ be the operator system defined by

$$S = \{p(e^{i\theta}) + \overline{q(e^{i\theta})} : p, q \text{ are polynomials}\} .$$

Then the map $\phi: S \to L(H)$ defined by $\phi(p + \overline{q}) = p(T) + q(T)^*$ is positive.

Proof. Let $\tau(e^{i\theta})$ be positive in S so that $\tau(e^{i\theta}) = \Sigma_{\ell,k=-N}^{+N} \alpha_\ell \overline{\alpha}_k e^{i(\ell-k)\theta}$. We must prove that

$$\phi(\tau) = \Sigma_{\ell,k=-n}^{+n} \alpha_\ell \overline{\alpha}_k T(\ell-k)$$

is a positive operator, where we define

$$T(j) = \begin{cases} T^j , & j \geq 0 \\ T^{*-j} , & j < 0 \end{cases} .$$

To this end, fix a vector x in our Hilbert space H and note that

$$(*) <\phi(\tau)x,x> = \left\langle \begin{bmatrix} IT^* & \cdots & T^{*n} \\ T & & \\ \vdots & \ddots & \vdots \\ & & T^* \\ T^n & \cdots T & I \end{bmatrix} \begin{bmatrix} \overline{\alpha}_1 x \\ \vdots \\ \overline{\alpha}_n x \end{bmatrix} , \begin{bmatrix} \overline{\alpha}_1 x \\ \vdots \\ \overline{\alpha}_n x \end{bmatrix} \right\rangle ,$$

where the matrix operator on the right is acting on the direct sum of n copies of the Hilbert space H^n .

Thus, if we can show that the matrix operator is positive, we will be done. To this end, set

13

$$
R \;=\; \begin{bmatrix} 0 & . & . & . & . & . & 0 \\ & . & & & & & . \\ T & . & & & & & . \\ & & . & & & & . \\ 0 & . & & . & & & . \\ . & & & & . & & . \\ 0 & . & 0 & & T & & 0 \end{bmatrix} \;,
$$

and note that $R^{n+1} = 0$, $\|R\| \le 1$.

Using I to also denote the identity operator on H^n, we see that the matrix operator in $(*)$ can be written as,

$$
I + R + R^2 + \ldots + R^n + R^* + \ldots + R^{*n} = (I - R)^{-1} + (I - R^*)^{-1} - I .
$$

To see that this latter operator is positive, fix h in H^n and let $h = (I - R)y$ for y in H^n. One obtains

$$
\langle ((I - R)^{-1} + (I - R^*)^{-1} - I)h, h \rangle
$$

$$
= \langle y, (I - R)y \rangle + \langle (I - R)y, y \rangle - \langle (I - R)y, (I - R)y \rangle
$$

$$
= \|y\|^2 - \|Ry\|^2 \ge 0 ,
$$

since R is a contraction. □

Corollary 2.7 (von Neumann's Inequality). Let T be an operator on a Hilbert space with $\|T\| \le 1$. Then for any polynomial p,

$$
\|p(T)\| \le \|p\| ,
$$

where $\|p\| = \sup_\theta |p(e^{i\theta})|$.

Proof. Since the polynomials are dense in the functions that are continuous on $\mathbb{D}^-$ and analytic on $\mathbb{D}$, and since every continuous

14

real-valued function on Π is the real part of such an analytic function, the operator system S defined in Theorem 2.6 is dense in $C(\Pi)$. By Proposition 2.1, the map ϕ is bounded and hence extends to $C(\Pi)$. Clearly the extension will also be positive. Hence, ϕ is contractive by Theorem 2.4, from which the result follows.   □

We let $A(\mathbb{D})$ denote the functions that are analytic on $\mathbb{D}$ and continuous on $\mathbb{D}^-$. Clearly, $A(\mathbb{D})$ is a closed subalgebra of $C(\Pi)$. Since the polynomials are dense in $A(\mathbb{D})$, the above inequality guarantees that the homomorphism $p \to p(T)$ extends to a homomorphism of $A(\mathbb{D})$ and we denote the image of an element f simply by $f(T)$, so that one has $\|f(T)\| \le \|f\|$ for all f in $A(\mathbb{D})$.

Another consequence of Theorem 2.6 which we shall frequently use is that if a is an element of some unital C^*-algebra A , $\|a\| \le 1$, then there is a unital, positive map $\phi: C(\Pi) \to A$ with $\phi(p) = p(a)$. This observation is used in the following two results.

Corollary 2.8. Let B , C be C^*-algebras with unit, let A be a subalgebra of B , $1 \in A$, and let $S = A + A^*$. If $\phi: S \to C$ is positive, then $\|\phi(a)\| \le \|\phi(1)\| \cdot \|a\|$ for all a in A .

Proof. Let a be in A , $\|a\| \le 1$. By Proposition 2.1, we may extend ϕ to a positive map on the closure of S , S^- . As we remarked above, there is a positive map $\psi: C(\Pi) \to B$ with $\psi(p) = p(a)$. Since A is an algebra, the range of ψ is actually contained in S^- .

Clearly, the composition of positive maps is positive, so by Theorem 2.4,

$$\| \phi(a) \| = \| \phi \circ \psi(e^{i\theta}) \| \leq \| \phi \circ \psi(1) \| \cdot \| e^{i\theta} \| = \| \phi(1) \| \, .$$

Thus, ϕ is a contraction.   □

It is somewhat surprising that ϕ need not be a contraction on all of S. We shall see an example of this phenomenon in Chapter 5.

Corollary 2.9 (Russo-Dye). Let A and B be C^*-algebras with unit and let $\phi: A \to B$ be a unital, positive map. Then $\| \phi \| = \| \phi(1) \|$.

Proof. Apply Corollary 2.8.   □

So far we have concentrated on positive maps without indicating how positive maps arise. We close this section with two such results.

Lemma 2.10. Let A be a C^*-algebra, $S \subseteq A$ an operator system, and $f: S \to \mathbb{C}$ a linear functional with $f(1) = 1$, $\| f \| = 1$. If a is a normal element of A and $a \in S$, then $f(a)$ will lie in the closed convex hull of the spectrum of a .

Proof. Suppose not. Note that the convex hull of a compact set is the intersection of all closed discs containing the set. Thus, there will exist a λ and $r > 0$ such that $|f(a) - \lambda| > r$, while the spectrum of a , $\sigma(a)$ satisfies

$$\sigma(a) \subseteq \{z: |z - \lambda| \leq r\} \, .$$

But then $\sigma(a - \lambda \cdot 1) \subseteq \{z: |z| \leq r\}$, and since norm and spectral radius agree for normal elements, $\|a - \lambda 1\| \leq r$, while $|f(a - \lambda \cdot 1)| > r$. This contradiction completes the proof. □

Since the convex hull of the spectrum of a positive operator is contained in the non-negative reals, we see that Lemma 2.10 implies that such an f must be positive.

Unlike our previous results, we shall see (Exercise 2.9) that the hypothesis that our map is unital is crucial to the next two results.

<u>Proposition 2.11</u>. Let S be an operator system, B a unital C^*-algebra, and $\phi: S \rightarrow B$ a unital contraction. Then ϕ is positive.

<u>Proof</u>. Since B can be represented on a Hilbert space, we may, without loss of generality, assume that $B = L(H)$ for some Hilbert space H . Fix x in H , $\|x\| = 1$. Setting $f(a) = \langle \phi(a)x,x \rangle$, we have that $f(a) = 1$, $\|f\| \leq \|\phi\|$. By Lemma 2.10, if a is positive, then $f(a)$ is positive, and consequently $\phi(a)$ is positive. □

<u>Proposition 2.12</u>. Let A be a unital C^*-algebra and let M be a subspace of A containing 1 . If B is a unital C^*-algebra and $\phi: M \rightarrow B$ is a unital contraction, then ϕ extends uniquely to a positive map $\tilde{\phi}: M + M^* \rightarrow B$ with $\tilde{\phi}$ given by

$$\tilde{\phi}(a + b^*) = \phi(a) + \phi(b)^* .$$

Proof. If ϕ does extend to a positive map $\tilde{\phi}$, then by the self-adjointness of $\tilde{\phi}$ (Exercise 2.1), $\tilde{\phi}$ necessarily satisfies the above equation. So we must prove that this formula yields a well-defined, positive map.

Note that to prove that $\tilde{\phi}$ is well-defined, it is enough to prove that if a and a^* belong to M , then $\phi(a^*) = \phi(a)^*$. For this, set

$$S_1 = \{a: a \in M \text{ and } a^* \in M\} ,$$

then S_1 is an operator system and ϕ is a unital, contractive map on S_1 and hence positive by Proposition 2.11. Consequently, ϕ is self-adjoint on S_1 and so $\tilde{\phi}$ is well-defined.

To see that $\tilde{\phi}$ is positive, it is sufficient to assume that $B = L(H)$, fix x in H with $\|x\| = 1$, set $\tilde{\rho}(a) = \langle\tilde{\phi}(a)x,x\rangle$, and prove that $\tilde{\rho}$ is positive. Let $\rho: M \to \mathbb{C}$ be defined by $\rho(a) = \langle\phi(a)x,x\rangle$, then $\|\rho\| = 1$ and so by the Hahn-Banach Theorem, ρ extends to

$$\rho_1: M + M^* \to \mathbb{C} \text{ with } \|\rho_1\| = 1 .$$

But by Proposition 2.11, ρ_1 is positive and so,
$\rho_1\overline{(a + b^*)} = \rho(a) + \rho(b) = \tilde{\rho}(a + b^*)$. Hence, $\tilde{\rho}$ is positive. □

Note that the above result also shows that there is a unique, norm-preserving Hahn-Banach extension of ρ to $M + M^*$.

Example 2.13. A positive map need not have a positive extension unless the range is $\mathbb{C}$ (Exercise 2.10). Indeed, if the positive map of Example 2.2 had a positive extension to $C(\Pi)$, then by Corollary 1.9, this extension would be a contraction.

18

If S is an operator system, contained in a C^*-algebra A, and ϕ is a linear functional on S with $\phi(1) = 1$, then by Exercise 2.3 and Proposition 2.11, we see that ϕ is contractive if and only if ϕ is positive. These maps are called <u>states</u> on S.

There is a natural way that positive maps on operator systems arise in operator theory. Let X be a compact set in the complex plane and let $R(X)$ be the algebra of quotients of polynomials p/q, where the 0's of q lie off X. We shall regard $R(X)$ as a subalgebra of $C(\partial X)$, the continuous functions on the boundary of X. By the maximum modulus theorem, this endows $R(X)$ with the same norm as if we regarded it as a subalgebra of $C(X)$. We let $S = R(X) + \overline{R(X)}$. If T is in $L(H)$, with the spectrum of T, $\sigma(T)$, contained in X, then we have a well-defined homomorphism $\rho: R(X) \to L(H)$, defined by $\rho(p/q) = p(T)q(T)^{-1}$. If $\|\rho\| \leq 1$, then X is called a <u>spectral set for</u> T. This concept is partially motivated by von Neumann's inequality, which can be interpreted as saying that an operator T is a contraction if and only if the closed unit disk is a spectral set for T.

By Proposition 2.12, if X is a spectral set for T, then there is a well-defined, positive map $\tilde{\rho}: S \to L(H)$ given by $\tilde{\rho}(f + \bar{g}) = f(T) + g(T)^*$. Conversely, if the above map $\tilde{\rho}$ is well-defined and positive, then by Corollary 2.8, X is a spectral set for T.

2.14. <u>Non-Unital C^*-algebras</u>. Let A be a non-unital C^*-algebra. We wish to make some comments on positive maps in this case. Let B be a C^*-algebra and let $\phi: A \to B$ be a positive map. We claim that ϕ is automatically bounded. To see this, note that it is enough to prove that ϕ is bounded on A^+. To this end, suppose that ϕ is not bounded, then there would exist a sequence p_n in A^+, $\|p_n\| \leq 1$, with $\|\phi(p_n)\| \geq n^3$.

Let $p = \sum_n n^{-2} p_n$, then we have that $n^{-2} p_n \leq p$ and so,

$$n \leq \|\phi(n^{-2} p_n)\| \leq \|\phi(p)\|$$

for all n, an obvious contradiction. Thus, ϕ is bounded.

The second observation is that every non-unital C^*-algebra A embeds in a unital C^*-algebra, A_1 [34]. Furthermore, positive maps on A extend to positive maps on A_1.

To see this second statement, first note that A is a closed, 2-sided ideal in A_1, and so the map $a + \lambda 1 \to \lambda$, for a in A, is a *-homomorphism. Thus, if $a + \lambda \cdot 1$ is positive, then $\lambda \geq 0$. Now, let $\phi: A \to B$ be positive, then since ϕ is bounded, there exists q in B such that $\phi(p) \leq q$ for all positive p, with $\|p\| \leq 1$. Define $\phi_1: A_1 \to B$ by $\phi_1(a + \lambda 1) = \phi(a) + \lambda q$. This map is positive, since if $a + \lambda 1 \geq 0$, then $-\lambda^{-1} a \leq 1$, so that $\phi(-\lambda^{-1} a) \leq q$ or $0 \leq \phi(a) + \lambda q$.

NOTES

For a survey of the theory of positive maps, see [105].

The idea of using von Neumann's inequality to prove the Russo-Dye result seems to have originated with Choi. The usual proof involves another important result of Russo-Dye. Namely, that the extreme points of the unit ball of a unital C^*-algebra are the unitary elements [93] (See [89] for an elegant proof of this result).

Lemma 2.10 is a minor adaptation of [5].

Von Neumann's original proof of his inequality [82] first showed that the inequality was met for the Möbius transformations of the disk, and then reduced verifying the inequality for a general analytic function to this special case. A later proof by Heinz [50] is based on the classical Cauchy-Poisson formula. Most other presentations rely on Sz.-Nagy's dilation theorem (Theorem 3.3). The Fejer-Riesz lemma can be found in [38].

Foias [41] has shown how particular von Neumann's inequality is to the theory of Hilbert spaces. He proves that if a Banach space has the property that every contraction operator on the Banach space satisfies Von Neumann's inequality, then that space is necessarily a Hilbert space.

Results of Ando [2] and of Sz.-Nagy-Foias [114] imply that von Neumann's inequality holds for pairs of commuting contractions, that is,

$$p(T_1,T_2) \leq \sup \{|p(z_1,z_2)| : |z_i| \leq 1 , \quad i = 1, 2\} ,$$

where T_1 and T_2 are commuting contractions and p is an arbitrary polynomial in 2 variables. Thus, by Proposition 2.12, there is a positive map,

$$\phi(p + \bar{q}) = p(T_1,T_2) + q(T_1,T_2)^* .$$

It would be interesting to know if a proof of this latter fact could be given along the lines of Theorem 2.6. Such a proof could perhaps shed some additional light on the rather paradoxical results of Crabbe-Davies [31] and of Varopoulos [119], that for 3 or more commuting contractions, the analogue of von Neumann's inequality fails.

EXERCISES

2.1 Let S be an operator system, B a C^*-algebra, and $\phi: S \to B$ a positive map. Prove that ϕ is self-adjoint, i.e., that $\phi(x^*) = \phi(x)^*$.

2.2 Let S be an operator system and $\phi: S \rightarrow B$ positive. Prove that ϕ
extends to a positive map on the norm closure of S .

2.3 Let S be an operator system and let $\phi: S \rightarrow \mathbb{C}$ be positive. Prove
that $\|\phi\| \leq \phi(1)$. [Hint: Given a , choose λ , $|\lambda| = 1$, such that
$|\phi(a)| = \phi(\lambda a)$].

2.4 Let S be an operator system and let $\phi: S \rightarrow \mathbb{C}(X)$ where $C(X)$
denotes the continuous functions on a compact Hausdorff space X .
Prove that if ϕ is positive, then $\|\phi\| \leq \|\phi(1)\|$.

2.5 (Schwarz Inequality) Let A be a C^* -algebra and let $\phi: A \rightarrow \mathbb{C}$ be a
positive linear functional. Prove that $|\phi(x^*y)|^2 \leq \phi(x^*x)\phi(y^*y)$.

2.6 Let T be an operator on a Hilbert space H , the numerical radius
of T is defined by $w(T) = \sup \{|<Tx,x>| : x \in H , \|x\| = 1\}$. Prove
that if $\phi: S \rightarrow L(H)$ is positive and $\phi(1) = 1$, then $w(\phi(a)) \leq \|a\|$.

2.7 Let T be an operator on a Hilbert space. Prove that $w(T) \leq 1$ if
and only if $2 + (\lambda T) + (\lambda T)^* \geq 0$ for all complex numbers λ with
$|\lambda| = 1$.

2.8 Prove that $w(T)$ defines a norm on $L(H)$, with $w(T) \leq \|T\| \leq 2w(T)$.
Show that both inequalities are sharp.

2.9 Let S be an operator system, B a C^* -algebra and $\phi: S \rightarrow B$ a linear
map such that $\phi(1)$ is positive, $\|\phi(1)\| = \|\phi\|$. Give an example to
show that ϕ need not be positive. In a similar vein, show that if
M is as in Proposition 1.12, $\phi(1)$ is positive, with $\|\phi(1)\| = \|\phi\|$,
then $\tilde{\phi}$ need not be well-defined.

2.10 (Krein) Let S be an operator system contained in the C^*-algebra A, and let $\phi: S \to \mathbb{C}$ be positive. Prove that ϕ can be extended to a positive map on A.

2.11 Prove that for the following element of M_{n+1},

$$\left\| \begin{bmatrix} a_0 & 0 & \cdots & 0 \\ a_1 & a_0 & & \\ & & \ddots & \\ & & & 0 \\ a_n & \cdots & a_1 & a_0 \end{bmatrix} \right\| \leq \inf_{r(z)} \{ \| a_0 + a_1 z + \cdots + a_n z^n + r(z) \| \} ,$$

where $r(z)$ is a polynomial whose lowest degree term is strictly greater than n, and the latter norm is the supremum norm over the unit disk.

2.12 (Jorgensen) Show that $C = \{ p^2(t) : p \text{ a polynomial} \}$ is dense in $C([0,1])^+$. Let $S = \{ p + \bar{q} : p, q \text{ are polynomials} \}$, and set $\phi(p + \bar{q}) = p(2) + q(2)$. Show that ϕ is positive on C, defined on a dense subset S of $C([0,1])$, but still does not extend to a positive map on $C([0,1])$. Compare with Proposition 2.1 and Exercise 2.2.

2.13 Let X be a compact subset of $\mathbb{C}$. Prove that if X is a subset of $\mathbb{R}$, then X is a spectral set for T if and only if $T = T^*$ and $\sigma(T)$ is contained in X. Prove that if X is contained in the unit circle, then X is a spectral set for T if and only if T is a unitary with $\sigma(T)$ contained in X.

2.14 (von Neumann) Let X be a compact subset of $\mathbb{C}$, with $R(X)$ dense in $C(\partial X)$. Prove that X is a spectral set for T if and only if T is normal and $\sigma(T)$ is contained in ∂X .

2.15 In this exercise we give an alternative proof of von Neumann's inequality. Let $T \in L(H)$ with $\|T\| < 1$ and let p and q denote arbitrary polynomials.

i) Let $P(t;T) = (1 - e^{-it}T)^{-1} + (1 - e^{it}T^*)^{-1} - 1$, and show that $P(t;T) \geq 0$ for all t .

ii) Show that $p(T) + q(T)^* = \frac{1}{2\pi} \int_0^{2\pi} (p(e^{it}) + \overline{q(e^{it})}) P(t;T) \, dt.$

iii) Deduce von Neumann's inequality.

3 Completely positive maps

Let A be a C^*-algebra and let M be a subspace, then we shall call M an operator space. Clearly, $M_n(M)$ can be regarded as a subspace of $M_n(A)$, and we let $M_n(M)$ have the norm structure that it inherits from the (unique) norm structure on the C^*-algebra $M_n(A)$. We make no attempt at this time to define a norm structure on $M_n(M)$ without reference to A. Similarly, if $S \subseteq A$ is an operator system, then we endow $M_n(S)$ with the norm and order structure that it inherits as a subspace of $M_n(A)$.

As before, if B is a C^*-algebra and $\phi: S \to B$ is a linear map. then we define $\phi_n: M_n(S) \to M_n(B)$ by $\phi_n((a_{i,j})) = (\phi(a_{i,j}))$. We call ϕ n-positive if ϕ_n is positive and we call ϕ completely positive if ϕ is n-positive for all n. We call ϕ completely bounded if $\sup_n \|\phi_n\|$ is finite and set

$$\|\phi\|_{cb} = \sup_n \|\phi_n\| .$$

Note that $\|\cdot\|_{cb}$ is a norm on the space of completely bounded maps. We use the terms completely isometric and completely contractive to indicate that each ϕ_n is isometric and that $\|\phi\|_{cb} \leq 1$, respectively. We note that if ϕ is n-positive, then ϕ is k-positive for $k \leq n$. Also, $\|\phi_k\| \leq \|\phi_n\|$ for $k \leq n$ (Exercise 3.1).

In this section we investigate some of the elementary properties of these classes of maps. We begin by relating some of the results of the previous section to these concepts.

<u>Lemma 3.1.</u> Let A be a C^*-algebra with unit and let a belong to A .
Then $\|a\| \leq 1$ if and only if

$$\begin{bmatrix} 1 & a \\ a^* & 1 \end{bmatrix}$$

is positive.

<u>Proof.</u> We represent A on a Hilbert space H via $\pi: A \to L(H)$ and
set $A = \pi(a)$.

If $\|A\| \leq 1$, then for any vectors x , y in H ,

$$\left\langle \begin{bmatrix} I & A \\ A^* & I \end{bmatrix} \begin{bmatrix} x \\ y \end{bmatrix}, \begin{bmatrix} x \\ y \end{bmatrix} \right\rangle$$

$$= \langle x,x \rangle + \langle Ay,x \rangle + \langle x,Ay \rangle + \langle y,y \rangle \geq \|x\|^2 - 2\|A\| \|y\| \|x\| + \|y\|^2 \geq 0 .$$

Conversely, if $\|A\| > 1$, then there exist unit vectors x and y
such that $\langle Ay,x \rangle < -1$ and the above inner product will be negative. □

<u>Proposition 3.2.</u> Let S be an operator system, B a C^*-algebra with
unit, and $\phi: S \to B$ a unital, 2-positive map. Then ϕ is contractive.

<u>Proof.</u> Let $a \in S$, $\|a\| \leq 1$, then

$$\phi_2 \begin{bmatrix} 1 & a \\ a^* & 1 \end{bmatrix} = \begin{bmatrix} 1 & \phi(a) \\ \phi(a)^* & 1 \end{bmatrix}$$

is positive and hence $\|\phi(a)\| \leq 1$. □

26

Proposition 3.3. Let A and B be C^*-algebras with unit, let M be a subspace of A, $1 \in M$, and let $S = M + M^*$. If $\phi: M \rightarrow B$ is unital and 2-contractive (i.e., $\|\phi_2\| \leq 1$), then the map $\tilde{\phi}: S \rightarrow B$ given by $\tilde{\phi}(a + b^*) = \phi(a) + \phi(b)^*$ is 2-positive and contractive.

Proof. Since ϕ is contractive, $\tilde{\phi}$ is well-defined by Proposition 2.12. Note that $M_2(S) = M_2(M) + M_2(M)^*$ and that $(\tilde{\phi})_2 = (\widetilde{\phi_2})$. Since ϕ_2 is contractive, again by Proposition 2.12, $\widetilde{\phi_2}$ is positive and so $\tilde{\phi}$ is contractive by Proposition 3.2. □

Proposition 3.4. Let A and B be C^*-algebras with unit, let M be a subspace of A, $1 \in M$, and let $S = M + M^*$. If $\phi: M \rightarrow B$ is unital and completely contractive, then $\tilde{\phi}: S \rightarrow B$ is completely positive and completely contractive.

Proof. We have that $\tilde{\phi}_n$ is positive since ϕ_n is unital and contractive and $\tilde{\phi}_n$ is contractive since $\tilde{\phi}_{2n} = (\tilde{\phi}_n)_2$ is positive. □

We've glossed over one point in the above proof. Namely, we've identified $M_{2n}(S)$ with $M_2(M_n(S))$. It's obvious how to do this, one simply "erases" the additional brackets in an element of $M_2(M_n(S))$. One must check, however, that the norms one defines are the same in each instance. One has that $M_2(M_n(S))$ inherits its norm from $M_2(M_n(A))$, while $M_{2n}(S)$ inherits its norm from $M_{2n}(A)$. However, this "erasure" operation defines a *-isomorphism between $M_2(M_n(A))$ and $M_{2n}(A)$, thus the norms are indeed the same.

Now that we've seen some of the advantages of considering maps in these two classes, let's begin by describing some maps that belong to these two classes. Let A and B be C^*-algebras. First, note that if $\pi: A \to B$ is a *-homomorphism, then π is completely positive and completely contractive, since each map, $\pi_n: M_n(A) \to M_n(B)$, is a *-homomorphism, and *-homomorphisms are both positive and contractive. For a second class of maps, fix x and y in A and define $\phi: A \to A$ by $\phi(a) = x\,a\,y$. Note that if $(a_{i,j})$ is in $M_n(A)$, then

$$
\|\phi_n(a_{i,j})\| = \|(x\,a_{i,j}\,y)\| = \left\| \begin{bmatrix} x & 0 \ldots 0 \\ 0 & \cdot \;\; \cdot \;\; \vdots \\ \vdots & \cdot \;\; \cdot \;\; 0 \\ 0 \ldots 0 & x \end{bmatrix} \begin{bmatrix} a_{11} \cdot \cdot a_{1n} \\ \vdots \qquad \vdots \\ a_{n1} \cdot \cdot a_{nn} \end{bmatrix} \begin{bmatrix} y & 0 \ldots 0 \\ 0 & \cdot \;\; \cdot \;\; \vdots \\ \vdots & \cdot \;\; \cdot \;\; 0 \\ 0 \ldots 0 & y \end{bmatrix} \right\|
$$

$$
\leq \|x\| \cdot \|(a_{i,j})\| \cdot \|y\| .
$$

Thus, ϕ is completely bounded and $\|\phi\|_{cb} \leq \|x\| \cdot \|y\|$. A similar calculation shows that if $x = y^*$, then ϕ is completely positive.

Combining these two examples we obtain the proto-typical example. Namely, let H_1 and H_2 be Hilbert spaces, let $v_i: H_1 \to H_2$, $i = 1, 2$, be bounded operators, and let $\pi: A \to L(H_2)$ be a *-homomorphism. Define a map $\phi: A \to L(H_1)$ via $\phi(a) = v_2^*\pi(a)v_1$, then ϕ is completely bounded with $\|\phi\|_{cb} \leq \|v_1\| \cdot \|v_2\|$ and if $v_1 = v_2$, then ϕ is completely positive. We shall prove in Chapter 8 that all completely bounded maps have this form.

In each of the above examples, we see that the completely positive maps are all completely bounded. This is always the case.

Proposition 3.5. Let $S \subseteq A$ be an operator system, let B be a C^*-algebra, and let $\phi: S \to B$ be completely positive. Then ϕ is completely bounded and $\|\phi(1)\| = \|\phi\| = \|\phi\|_{cb}$.

Proof. Clearly, we have that $\|\phi(1)\| \leq \|\phi\| \leq \|\phi\|_{cb}$, so it is sufficient to show $\|\phi\|_{cb} \leq \|\phi(1)\|$. To this end, let $A = (a_{i,j})$ be in $M_n(S)$ with $\|A\| \leq 1$, and let I_n be the unit of $M_n(A)$, i.e., the diagonal matrix with 1's on the diagonal. Since

$$\begin{bmatrix} I_n & A \\ A^\star & I_n \end{bmatrix}$$

is positive, we have that

$$\phi_{2n}\left(\begin{bmatrix} I_n & A \\ A^\star & I_n \end{bmatrix}\right) = \begin{bmatrix} \phi_n(I_n) & \phi_n(A) \\ \phi_n(A)^\star & \phi_n(I_n) \end{bmatrix}$$

is positive. Thus, by Exercise 3.2ii, $\|\phi_n(A)\| \leq \|\phi_n(I_n)\| = \|\phi(1)\|$, which completes the proof. □

3.6. Schur Products. As an application of the above result, we study the Schur product of matrices. If $A = (a_{i,j})$, $B = (b_{i,j})$ are elements of M_n , then we define the Schur product by

$$A * B = (a_{i,j} \cdot b_{i,j}) .$$

For fixed A , this gives rise to a linear map,

$$S_A: M_n \rightarrow M_n , \quad \text{via} \quad S_A(B) = A * B .$$

In order to study this map, we need to recall a few facts about tensor products. Let A be in M_n and B be in M_m so that A and B can be thought of as linear transformations on $\mathbb{C}^n$ and $\mathbb{C}^m$, respectively. Then $A \otimes B$ is the linear transformation on $\mathbb{C}^n \otimes \mathbb{C}^m \simeq \mathbb{C}^{nm}$, which is defined by setting $A \otimes B(x \otimes y) = Ax \otimes By$ and extending linearly. Let $\{e_1,\ldots, e_n\}$ and $\{f_1,\ldots, f_m\}$ be the canonical orthonormal bases for $\mathbb{C}^n$ and $\mathbb{C}^m$, respectively. If we order our basis for $\mathbb{C}^{nm}$ by $e_1 \otimes f_1$, $e_1 \otimes f_2$, $\ldots$, $e_1 \otimes f_m$, $e_2 \otimes f_1$, $\ldots$, $e_m \otimes f_m$, then the matrix for $A \otimes B$ with respect to this basis is given in block form by

$$\begin{bmatrix} a_{11}B & \cdots & a_{1n}B \\ \vdots & & \vdots \\ a_{n1}B & \cdots & a_{nn}B \end{bmatrix} .$$

Writing $A \otimes B = (A \otimes I)(I \otimes B)$, it is easy to see that $\|A \otimes B\| = \|A\| \cdot \|B\|$. Now let A and B be in M_n and define an isometry $V: \mathbb{C}^n \to \mathbb{C}^n \otimes \mathbb{C}^n$ by $V(e_i) = e_i \otimes e_i$. A simple calculation shows that

$$V^*(A \otimes B)V = A * B .$$

To see this, note that

$$\langle V^*(A \otimes B)Ve_j, e_i \rangle = \langle A \otimes B(e_j \otimes e_j), (e_i \otimes e_i) \rangle$$

$$= \langle Ae_j, e_i \rangle \cdot \langle Be_j, e_i \rangle = a_{i,j} \cdot b_{i,j} = \langle A * Be_j, e_i \rangle .$$

Thus, $\|S_A(B)\| = \|V^*(A \otimes B)V\| \leq \|A\| \cdot \|B\|$ and so $\|S_A\| \leq \|A\|$.

Similarly, if $(B_{i,j})$ is in $M_k(M_n)$, then

$(S_A)_k((B_{i,j})) = (V^*(A \otimes B_{i,j})V) =$

$$\begin{bmatrix} V^* & 0 & \cdots & 0 \\ 0 & \ddots & & \vdots \\ \vdots & & \ddots & 0 \\ 0 & \cdots & 0 & V^* \end{bmatrix} A \otimes \begin{bmatrix} B_{11} & \cdots & B_{1n} \\ \vdots & & \vdots \\ B_{n1} & \cdots & B_{nn} \end{bmatrix} \begin{bmatrix} V & 0 & \cdots & 0 \\ 0 & \ddots & & \vdots \\ \vdots & & \ddots & 0 \\ 0 & \cdots & 0 & V \end{bmatrix}$$

and so, $\|(S_A)_k\| \le \|A\|$ also.

It is easy to check that the tensor-product of positive operators is positive. This implies, in particular, that the Schur product of positive matrices is positive, and more generally, that if A is positive, then S_A is completely positive. Hence, for positive matrices we can obtain $\|S_A\|$ explicitly, by invoking Proposition 3.5. Namely, $\|S_A\| = \|S_A(I)\| =$
max $\{a_{i,i}: i = 1, \ldots, n\}$.

It is more difficult to calculate $\|S_A\|_{cb}$ when A is not positive. We return to this topic in Chapter 8. Clearly, if one decomposes
$A = (P_1 - P_2) + i(P_3 - P_4)$ with P_i positive, then

$$S_A = (S_{P_1} - S_{P_2}) + i(S_{P_3} - S_{P_4}) .$$ Thus,

$\|S_A\|_{cb} \le \|S_{P_1}\|_{cb} + \|S_{P_2}\|_{cb} + \|S_{P_3}\|_{cb} + \|S_{P_4}\|_{cb}$ and each of the right

hand terms is given by the maximum diagonal entry of the corresponding matrix. However, we shall see that this estimate can be far from $\|S_A\|_{cb}$.

The next result shows that for linear functionals, the adverb "completely" introduces nothing new.

Proposition 3.7. Let S be an operator space and let $f: S \to \mathbb{C}$ be a bounded linear functional. Then $\|f\|_{cb} = \|f\|$. Furthermore, if S is an operator system and f is positive, then f is completely positive.

Proof. Let $(a_{i,j})$ be in $M_n(S)$ and let $x = (x_1, \ldots, x_n)$, $y = (y_1, \ldots, y_n)$ be unit vectors in $\mathbb{C}^n$. We have that
$|<f_n(a_{i,j})x, y>| = |\Sigma_{i,j} f(a_{i,j})x_j\bar{y}_i| = |f(\Sigma_{i,j} a_{i,j}x_j\bar{y}_i)| \le$
$\|f\| \cdot \|\Sigma_{i,j} a_{i,j}x_j\bar{y}_i\|$. Thus, we must show that this latter element has

31

norm less than $\|(a_{i,j})\|$. To see this, note that the above sum is the
(1,1)-entry of the product

$$
\begin{bmatrix} \bar{y}_1 \cdot 1 \ldots \ldots \bar{y}_n \cdot 1 \\ 0 \ldots \ldots 0 \\ \vdots \qquad\qquad \vdots \\ 0 \ldots \ldots 0 \end{bmatrix}
\begin{bmatrix} a_{11} \cdots \ldots a_{1n} \\ \vdots \qquad\qquad \vdots \\ \vdots \qquad\qquad \vdots \\ a_{n1} \cdots \ldots a_{nn} \end{bmatrix}
\begin{bmatrix} x_1 \cdot 1 & 0 \ldots 0 \\ \vdots & \vdots \\ \vdots & \vdots \\ x_n \cdot 1 & 0 \ldots 0 \end{bmatrix} ,
$$

and that the outer two factors each have norm equal to one, since x and
y were chosen to be unit vectors.

To prove that f is completely positive whenever f is positive
reduces to showing that $\langle f_n(a_{i,j})x, x\rangle = f(\Sigma_{i,j}\, a_{i,j}x_j\bar{x}_i)$ is positive
whenever $(a_{i,j})$ is positive.

But using the above product with $x = y$, we see that the summation
that f is being evaluated at is the (1,1)-entry of a positive element
and hence is positive.   □

Let X be a compact Hausdorff space and let $C(X)$ be the C^*-algebra
of continuous functions on X . Note that every element $F = (f_{i,j})$ of
$M_n(C(X))$ can be thought of as a continuous matrix-valued function and that
multiplication and the *-operation in $M_n(C(X))$ are just the pointwise
multiplication and *-operation of the matrix-valued functions. Thus, one
way to make $M_n(C(X))$ into a C^*-algebra is to set $\|F\| = \sup\,\{\|F(x)\|: x \in X\}$
and by the uniqueness of C^*-norms, this is the only way. With these
observations, the following is a direct consequence of Proposition 3.7.

Theorem 3.8. Let S be an operator space and let $\phi: S \to C(X)$ be a
bounded linear map. Then $\|\phi\|_{cb} = \|\phi\|$. Furthermore, if S is an operator

system and ϕ is positive, then ϕ is completely positive.

Proof. Let $x \in X$, and define $\phi^x: S \to \mathbb{C}$ by $\phi^x(a) = \phi(a)(x)$.
By the above observations, $\|\phi_n\| = \sup \{\|\phi_n^x\|: x \in X\} = \sup \{\|\phi^x\|: x \in X\} = \|\phi\|$.
Similarly, $\phi_n((a_{i,j}))$ is positive if and only if $\phi_n^x((a_{i,j}))$ is positive
for all $x \in X$, from which the second statement follows.   □

A commutative domain is also enough to ensure that positive maps are
completely positive. However, we shall see that it is not enough to
guarantee that bounded maps are completely bounded.

Lemma 3.9. Let $(p_{i,j})$ be a positive matrix and let q be a positive
element of some C^* -algebra B . Then $(q \cdot p_{i,j})$ is positive in $M_n(B)$.

Proof. Straightforward.   □

Theorem 3.10 (Stinespring). Let B be a C^* -algebra and let
$\phi: C(X) \to B$ be positive. Then ϕ is completely positive.

Proof. Let $P(x)$ be positive in $M_n(C(X))$, then given $\varepsilon > 0$ and
arguing as in Theorem 2.4, we obtain a partition of unity $\{u_\ell(x)\}$ and
positive matrices $P_\ell = (p_{i,j}^\ell)$ such that $\|P(x) - \Sigma_\ell u_\ell(x)P_\ell\| < \varepsilon$. But
$\phi_n(u_\ell \cdot P_\ell) = \phi_n((u_\ell \cdot p_{i,j}^\ell)) = (\phi(u_\ell) \cdot p_{i,j}^\ell)$, which is positive by
Lemma 3.9. Thus, $\phi_n(P)$, to within ε , is a sum of positive elements
and hence is positive.   □

Just as for a commutative domain or range positivity guarantees complete positivity, we shall see that if the domain or range isn't too badly noncommutative, then slightly weaker hypotheses guarantees complete positivity. For now we restrict the domain.

Lemma 3.11. Let A be a C^*-algebra. Then every positive element of $M_n(A)$ is a sum of n positive elements of the form $(a_i^* a_j)$ for some $\{a_1, \ldots, a_n\} \subseteq A$.

Proof. We remark that if we let R be the element of $M_n(A)$ whose k-th row is $a_1, \ldots, a_n$ and whose other entries are 0, then $R^*R = (a_i^* a_j)$, so such an element is positive. Now let P be positive, so $P = B^*B$ and write $B = R_1 + \ldots + R_n$, where R_k is the k-th row of B and 0 elsewhere.

We have that $P = B^*B = R_1^*R_1 + \ldots + R_n^*R_n$, since $R_i^*R_j = 0$ when $i \neq j$. □

We note that by the above lemma, to verify that $\phi: A \to B$ is n-positive it is sufficient to check that $(\phi(a_i^* a_j))$ is positive for all $\{a_1, \ldots, a_n\}$.

Theorem 3.12 (Choi). Let B be a C^*-algebra, let $\phi: M_n \to B$, and let $\{E_{i,j}\}_{i,j=1}^n$ denote the standard matrix units for M_n. The following are equivalent:

  i) ϕ is completely positive.

  ii) ϕ is n-positive.

  iii) $(\phi(E_{i,j}))_{i,j=1}^n$ is positive in $M_n(B)$.

34

Proof. Clearly, i) implies ii) , and since $(E_{i,j})_{i,j=1}^{n}$ is positive, ii) implies iii) . Thus, we shall prove that iii) implies i) .

For this it is sufficient to assume that $B = L(H)$. Fix k and let $x_1, \ldots, x_k$ belong to H , $B_1, \ldots, B_k$ belong to M_n . By the above, it is sufficient to prove that $\Sigma_{i,j} \langle \phi(B_i^*B_j)x_j, x_i \rangle$ is positive.

Write $B_\ell = \Sigma_{r,s=1}^{n} b_{r,s,\ell} E_{r,s}$ so that

$$B_i^*B_j = \Sigma_{r,s,t=1}^{n} \overline{b_{r,s,i}} b_{r,t,j} E_{s,t} .$$

Set $y_{t,r} = \Sigma_{j=1}^{k} b_{r,t,j} x_j$, then $\Sigma_{i,j} \langle \phi(B_i^*B_j)x_j x_i \rangle =$

$\Sigma_{r=1}^{n} \Sigma_{s,t=1}^{n} \langle \phi(E_{s,t})(\Sigma_{i,j} \overline{b_{r,s,i}} b_{r,t,j} x_j), x_i \rangle = \Sigma_{r=1}^{n} \Sigma_{s,t} \langle \phi(E_{s,t}) y_{t,r}, y_{s,r} \rangle .$

But for each r , this last sum is positive, since $(\phi(E_{s,t}))_{s,t=1}^{n}$ is positive. Thus, we've expressed our original sum as the sum of n positive quantities. This completes the proof. □

By combining the results of this chapter with the technique used in the proof of Theorem 2.6, some fairly deep operator theory results can be readily obtained. Recall that if T is in $L(H)$, then we define the numerical radius of T by

$$w(T) = \sup \{|\langle Tx,x \rangle|: \|x\| \leq 1 , x \in H\} .$$

The following result is essentially due to Berger [9].

Theorem 3.13. Let T be in $L(H)$ and let $S \subseteq C(\Pi)$ be the operator system defined by $S = \{p + \overline{q}: p, q \text{ polynomials}\}$. The following are equivalent:

35

i) $w(T) \le 1$.

ii) The map $\phi: S \to L(H)$ defined by

$$\phi(p + \bar{q}) = p(T) + q(T)^* + \overline{(p(0) + q(0))}I \quad \text{is positive.}$$

Proof. Let R_n be the $n \times n$ operator matrix whose subdiagonal entries are T and whose remaining entries are 0 . It is not difficult to show that $w(R_n) \le w(T)$ (Exercise 3.13).

Mimicking the first part of the proof of Theorem 2.6, we see that the above map ϕ is positive, provided that the operator matrices,

$$(*) \qquad \begin{bmatrix} 2 & T^* & \cdots & T^{*n} \\ T & \cdot & \cdot & \cdot \\ \cdot \cdot & \cdot & \cdot & \cdot \\ \cdot & \cdot & \cdot & \cdot T^* \\ \cdot & \cdot & \cdot & \cdot \\ T^n & \cdots T & \cdot 2 \end{bmatrix}$$

are positive for all n .

Note that as in Theorem 2.6, $R_n^{n+1} = 0$ and so the matrix $(*)$ can be written as $(I - R_n)^{-1} + (I - R_n^*)^{-1}$. Fix a vector $x = (I - R_n)y$ and compute

$$\langle ((I - R_n)^{-1} + (I - R_n^*)^{-1})x, x \rangle = 2\|y\|^2 - 2\mathrm{Re}\langle R_n y, y \rangle \ .$$

Thus, $(*)$ is positive if and only if $w(R_n) \le 1$ (Exercise 3.13 iv).

If $w(T) \le 1$, then $w(R_n) \le 1$ and so $(*)$ is positive, which implies that ϕ is positive.

Conversely, if ϕ is positive, then, since S is dense in $C(\Pi)$, ϕ will be completely positive by Theorem 3.10. Note that the matrix

$$(**)\quad \begin{bmatrix} 1 & \bar{z} & \cdots & \bar{z}^n \\ z & 1 & & \\ \vdots & & \ddots & \bar{z} \\ z^n & \cdots z & & 1 \end{bmatrix} = \begin{bmatrix} 1 & 0 & \cdots & 0 \\ 0 & z & & \vdots \\ \vdots & & \ddots & 0 \\ 0 & \cdots & 0 & z^n \end{bmatrix} \begin{bmatrix} 1 & \cdots & & 1 \\ \vdots & & & \vdots \\ \vdots & & & \vdots \\ 1 & \cdots & & 1 \end{bmatrix} \begin{bmatrix} 1 & 0 & \cdots & 0 \\ 0 & \bar{z} & & \vdots \\ \vdots & & \ddots & 0 \\ 0 & \cdots 0 & & \bar{z}^n \end{bmatrix}$$

is positive in $M_n(C(\Pi))$, and so its image under ϕ_n will be positive. But the image of (**) under ϕ_n is (*) . Thus, (*) is positive for all n and hence $w(R_n) \leq 1$.

Let $x \in H$, $\|x\| = 1$, and let $y = (x \oplus \ldots \oplus x)/\sqrt{n}$ be a unit vector in the direct sum of n copies of H . We have that

$$1 \geq |\langle R_n y, y\rangle| = \frac{n-1}{n} |\langle Tx, x\rangle| ,$$

from which it follows that $w(T) \leq \frac{n}{n-1}$ for all n .

Thus, $w(T) \leq 1$, which completes the proof of the theorem. □

Note that if $w(T) \leq 1$ and ϕ is as above, then since ϕ is positive, $\|\phi\| = \|\phi(1)\| = 2$. Thus, if p is a polynomial, then $\|p(T)\| = \|\phi(p) - p(0)I\| \leq 3\|p\|$. In particular, if $w(T) \leq 1$, then the functional calculus can be extended from polynomials to the disc algebra, $A(\mathbb{D})$.

Corollary 3.14 (Berger-Kato-Stampfli). Let T be in $L(H)$ with $w(T) \leq 1$ and let f be in $A(\mathbb{D})$ with $f(0) = 0$. Then $w(f(T)) \leq \|f\|$.

Proof. It is sufficient to assume that f is a polynomial and that $\|f\| \leq 1$. Let ϕ be the map of Theorem 3.13 for T . We must show that the map $\psi(p + \bar{q}) = p(f(T)) + q(f(T))^* + (p(0) + \overline{q(0)})I$ is positive.

But if $p + \bar{q}$ is positive, then $p \circ f + q \circ f$ is positive, and thus,

$$\psi(p + \bar{q}) = p(f(T)) + q(f(T))^* + (p(0) + \overline{q(0)})I = \phi(p \circ f + \overline{q \circ f})$$

is positive. □

Corollary 3.15 (Berger). Let T be in $L(H)$. Then $w(T^n) \leq w(T)^n$.

Proof. We may assume $w(T) = 1$, but then applying Corollary 3.14, with $f(z) = z^n$, yields the result. □

NOTES

Lemma 3.1 is an observation used in the work of Choi and Effros [22].

Proposition 3.4 appears in Arveson [3], which is the source for many of the applications of complete positivity to operator theory.

Theorem 3.10 is due to Stinespring [104], where the term completely positive is introduced and used. Stinespring's proof was measure theoretic.

Theorem 3.12 can be found in Arveson [6] and Choi [16] and [18].

Theorem 3.13, Corollary 3.14, and Corollary 3.15 can be found in Berger [9], Kato [65], and Berger-Stampfli [11]. For some related work see [10], [54], and [121]. These ideas were further generalized by the theory of C_p operators in Sz.-Nagy-Foias [114] (see Exercises 4.16 and 8.10). For an elementary proof of Corollary 3.15, see [88].

Further results in the direction of Exercise 3.10 can be found in Tomiyama's survey article [118].

Exercise 3.11 is an unpublished result of Smith [100].

Exercise 3.6 is a result of Choi [16].

EXERCISES

3.1 Prove that $\|\phi_n\| \leq \|\phi_k\|$ for $n \leq k$ and that if ϕ_k is positive, then ϕ_n is positive.

3.2 Let P, Q, A be operators on some Hilbert space H with P and Q positive.

(i) Show that
$$\begin{bmatrix} P & A \\ A^* & Q \end{bmatrix} \geq 0$$

if and only if $|<Ax,y>|^2 \leq <Py,y> \cdot <Qx,x>$ for all x, y in H.

(ii) Show that if
$$\begin{bmatrix} P & A \\ A^* & P \end{bmatrix} \geq 0,$$

then $A^*A \leq \|P\| \cdot P$ and in particular $\|A\| \leq \|P\|$.

3.3 Prove a non-unital version of Proposition 3.2.

3.4 (Schwarz Inequality for 2-positive maps) Let A and B be C^*-algebras, $\phi: A \to B$ 2-positive. Prove that
$\phi(a)^*\phi(a) \leq \|\phi(1)\|\phi(a^*a)$ and that $\|\phi(a^*b)\|^2 \leq \|\phi(a^*a)\| \cdot \|\phi(b^*b)\|$.

$$\left[\text{Hint: Consider } \begin{bmatrix} 1 & a \\ 0 & 0 \end{bmatrix} \text{ and } \begin{bmatrix} a & b \\ 0 & 0 \end{bmatrix} \right]$$

3.5 Let A be a C^*-algebra with unit. Show that the maps $\mathrm{Tr}, \sigma: M_n(A) \to A$ defined by $\mathrm{Tr}((a_{i,j})) = \Sigma_i\, a_{i,i}$, and $\sigma((a_{i,j})) = \Sigma_{i,j}\, a_{i,j}$ are completely positive maps. Deduce that if $\|(a_{i,j})\| \leq 1$, then $\|\Sigma_{i,j}\, a_{i,j}\| \leq n$.

3.6 (Choi) Let A be a C^*-algebra, let λ be a complex number with $|\lambda| = 1$, let U_λ be the unitary element of $M_n(A)$ that is diagonal with $u_{i,i} = \lambda^i$, and let Diag: $M_n(A) \to M_n(A)$ be defined by Diag $((a_{i,j})) = (b_{i,j})$, where $b_{i,j} = 0$, for $i \neq j$ and $b_{i,i} = a_{i,i}$.

(i) Show that $U_\lambda^*(a_{i,j})U_\lambda = (\lambda^{j-i}a_{i,j})$.

(ii) By considering the non-trivial n-th roots of unity, show the map $\Phi: M_n(A) \to M_n(A)$ defined by $\Phi(A) = n$ Diag $(A) - A$ is completely positive.

(iii) Show that the map $\Phi: M_n \to M_n$ defined by $\Phi(A) = (n - 1)$ Diag $(A) - A$ is not completely positive.

3.7 Let A and B be C^*-algebras with unit and let $\phi_1, \phi_2: A \to B$ be bounded linear maps with $\phi_1 \pm \phi_2$ completely positive. Prove that $\|\phi_2\|_{cb} \leq \|\phi_1(1)\|$.

3.8 Let A be a C^*-algebra with unit. Define $T_1, T_2: M_n(A) \to M_n(A)$ by $T_1((a_{i,j})) = (b_{i,j})$ where $b_{i,i} = \Sigma_\ell a_{\ell,\ell}, b_{i,j} = 0$, $i \neq j$ and $T_2((a_{i,j})) = (c_{i,j})$ where $c_{i,j} = a_{j,i}$.

(i) Fix k and ℓ, $k \neq \ell$, and define $U_{k,\ell}^{\pm}$ to be 1 in the (k,ℓ)-entry and ± 1 in the (ℓ,k)-entry. Show that $T_1(A) \pm T_2(A) = \frac{1}{2}\Sigma_{k \neq \ell} U_{k,\ell}^{\pm *} AU_{k,\ell}^{\pm}$.

(ii) Show that $T_1 \pm T_2$ are completely positive and that $\|T_2\|_{cb} \leq n$.

(iii) By considering $A = \mathbb{C}$, show that $T_2{}_{cb} = n$.

3.9 Let A be a C^*-algebra and let A^{op} denote the set A with the same norm and *-operation, but with a multiplication defined by $a \circ b = ba$.

(i) Prove that A^{op} is a C^*-algebra.

(ii) Prove that M_2 and M_2^{op} are *-isomorphic via the transpose map.

(iii) Show that the identity map from A to A^{op} is always positive.

(iv) Prove that the identity map from M_2 to M_2^{op} is not 2-positive.

(v) Prove that the identity map from A to A^{op} is completely positive if and only if A is commutative.

3.10 (Tomiyama) Let A and B be unital C^*-algebras, and let $(a_{i,j})$ be in $M_n(A)$.

(i) Prove that $\|(a_{i,j})\| \leq \|(\|a_{i,j}\|)\| \leq (\Sigma_{i,j} \|a_{i,j}\|^2)^{\frac{1}{2}} \leq n\|(a_{i,j})\|$, and give examples to show that all of these inequalities are sharp.

(ii) Let M be an operator space in A and let $\phi: M \to B$ be bounded. Prove that $\|\phi_n\| \leq n\|\phi\|$.

3.11 (Smith) Let M be an operator space and let $\phi: M \to M_n$ be bounded. Prove that $\|\phi\|_{cb} \leq n\|\phi\|$. [Hint: Write $\phi(a) = \Sigma_{i,j=1}^n \phi_{i,j}(a) \otimes E_{i,j}$, where $\phi_{i,j}: M \to \mathbb{C}$] .

3.12 Let X be a compact subset of $\mathbb{C}$ and let $R(X)$ be the quotients of polynomials with poles off X . We may regard $R(X)$ as a subalgebra of $C(\partial X)$ or of $C(X)$. Prove that with respect to these two embeddings, the identity map from $R(X)$ to $R(X)$ is completely isometric. Use this result to deduce that the real part of a function in $R(X)$ is positive on X if and only if it is positive on ∂X .

3.13 Let S_n denote the cyclic forward shift on $\mathbb{C}^n$. That is, $S_n e_j = e_{j+1}$ (mod n) , where $e_0, \ldots, e_{n-1}$ is the canonical basis for $\mathbb{C}^n$.

    (i)   Show that S_n is unitarily equivalent to a diagonal matrix whose entries are the n-th roots of unity.

    (ii) Let T be in $L(H)$. Show that $w(T) = w(T \otimes S_n)$.

    (iii) Let R_n be the $n \times n$ matrix operator whose subdiagonal entries are T and which is 0 elsewhere. Show that $w(R_n) \le w(T \otimes S_n)$. [Hint: Consider $x_\lambda = \lambda x_1 \oplus \ldots \oplus \lambda^n x_n$ with $|\lambda| = 1$] .

    (iv) Show that $\mathrm{Re} < R_n y, y > \; \le 1$ for all $\|y\| = 1$ if and only if $w(R_n) \le 1$.

3.14 In this exercise we outline an alternative proof of Theorem 3.13. Let $T \in L(H)$ with $w(T) < 1$ and let p and q be arbitrary polynomials.

    i)   Show directly that $\sigma(T)$ is contained in the open unit disk. [Hint: Recall Exercise 2.7.]

    ii) Let $Q(t;T) = (1 - e^{-it}T)^{-1} + (1 - e^{it}T^*)^{-1}$ and show that $Q(t;T) \ge 0$ for all t .

    iii) Show that
$$p(T) + q(T)^* + (p(0)+\overline{q(0)})I = \frac{1}{2\pi} \int_0^{2\pi} (p(e^{it}) + \overline{q(e^{it})}) \, Q(t;T) \, dt.$$

    iv) Deduce Theorem 3.13.

3.15 Let $\{A_n\}_{n=-\infty}^{+\infty}$ be a sequence in $L(H)$. Prove that $\phi(z^n) = A_n$ extends linearly to a completely positive map $\phi: C(\Pi) \to L(H)$ if and only if the $n \times n$ Toeplitz matrix (A_{i-j}) is positive for all n .

4 Dilation theorems

In general, a dilation theorem is a result which characterizes some class of maps into $L(H)$ as compressions to H of "nicer" maps into $L(K)$, where K is a Hilbert space containing H. One of the most general dilation theorems is Stinespring's theorem which characterizes completely positive maps from C^*-algebras into $L(H)$ as compressions of *-homomorphisms.

<u>Theorem 4.1 (Stinespring)</u>. Let A be a unital C^*-algebra and let $\phi: A \to L(H)$ be a completely positive map. Then there exists a Hilbert space K, a unital *-homomorphism $\pi: A \to L(K)$, and a bounded operator $V: H \to K$ with $\|\phi(1)\| = \|V\|^2$ such that

$$\phi(a) = V^*\pi(a)V .$$

<u>Proof</u>. Consider the algebraic tensor product $A \otimes H$ and define a symmetric bilinear function $< , >$ on this space by setting

$$<a \otimes x, b \otimes y> = <\phi(b^*a)x, y>_H$$

and extending linearly, where $< , >_H$ is the inner product on H.

The fact that ϕ is completely positive insures that $< , >$ is positive and semi-definite since

$$< \Sigma_{j=1}^n a_j \otimes x_j, \Sigma_{i=1}^n a_i \otimes x_i> = <\phi_n((a_i^*a_j)) \begin{bmatrix} x_1 \\ . \\ . \\ x_n \end{bmatrix}, \begin{bmatrix} x_1 \\ . \\ . \\ x_n \end{bmatrix}>_{H^n} \geq 0 ,$$

where $< , >_{H^n}$ denotes the inner product on the direct sum of n copies of H , H^n .

Positive semi-definite bilinear forms satisfy the Cauchy-Schwartz inequality,

$$|<u,v>|^2 \leq <u,u> \cdot <v,v> .$$

Thus, we have that

$$\{u \in A \otimes H| <u,u> = 0\} = \{u \in A \otimes H| <u,v> = 0 , \text{ for all } v \in A \otimes H\}$$

is a subspace, N of $A \otimes H$. The induced bilinear form on the quotient space $A \otimes H/N$ defined by

$$<u + N, v + N> = <u,v>$$

will be an inner product. We let K denote the Hilbert space that is the completion of the inner product space $A \otimes H/N$.

If $a \in A$, define a linear map $\pi(a): A \otimes H \to A \otimes H$ by

$$\pi(a)(\Sigma \; a_i \otimes x_i) = \Sigma \; (aa_i) \otimes x_i \; .$$

A matrix factorization shows that the following elements of $M_n(A)^+$ satisfy

$$(a_i^* a^* aa_j) \leq \|a^* a\| \cdot (a_i^* a_j) \; ,$$

and consequently,

$$<\pi(a)(\Sigma \; a_j \otimes x_j), \; \pi(a)(\Sigma \; a_i \otimes x_i)>$$

$$= \Sigma_{i,j} \; <\phi(a_i^* a^* aa_j)x_j, \; x_i>_H \leq \|a^* a\| \cdot \Sigma_{i,j} \; <\phi(a_i^* a_j)x_j, \; x_i>_H$$

$$= \|a\|^2 \cdot <\Sigma \; a_j \otimes x_j, \; \Sigma \; a_i \otimes x_i> \; .$$

Thus, $\pi(a)$ leaves N invariant and defines a linear transformation on $A \otimes H/N$, which we still denote by $\pi(a)$. The above inequality also shows that $\pi(a)$ is bounded with $\|\pi(a)\| \leq \|a\|$. Thus, $\pi(a)$ extends to a bounded linear operator on K, which we still denote by $\pi(a)$. It is straightforward to verify that the map $\pi: A \to L(K)$ is a unital *-homomorphism.

Now define $V: H \to K$ via

$$V(x) = 1 \otimes x + N,$$

then V is bounded, since

$$\|Vx\|^2 = <1 \otimes x, 1 \otimes x> = <\phi(1)x, x>_H \leq \|\phi(1)\| \cdot \|x\|^2.$$

Indeed, it is clear that $\|V\|^2 = \|\phi(1)\|$.

To complete the proof, we only need observe that

$$<V^*\pi(a)Vx, y>_H = <\pi(a) 1 \otimes x, 1 \otimes y> = <\phi(a)x, y>_H,$$

and so $V^*\pi(a)V = \phi(a)$. $\qquad\qquad\qquad\qquad$ □

There are several remarks to be made. First, note that if ϕ is unital, then V is an isometry. In this case we may identify H with the subspace VH of K. With this identification, V^* becomes the projection of K onto H, P_H. Thus, we see that

$$\phi(a) = P_H\pi(a)|_H.$$

So that when $\phi(1) = 1$, Stinespring's theorem is a dilation result in the sense described at the beginning of this section.

The second fact to note is that if H and A are separable, then the space K constructed above will be separable as well. Similarly, if H and A are finite dimensional, then K is finite dimensional.

The third remark is on the uniqueness of the Stinespring representation. We shall call a triple (π,V,K) a <u>Stinespring representation</u> for ϕ whenever $\phi(a) = V^*\pi(a)V$. Given a Stinespring representation (π,V,K) , let K_1 be the closed linear span of $\pi(A)VH$. It is easily verified that K_1 reduces $\pi(A)$ so that the restriction of π to K_1 defines a *-homomorphism, $\pi_1\colon A \to L(K_1)$.

Clearly, $VH \subseteq K_1$, so we have that $\phi(a) = V^*\pi_1(a)V$, i.e., that (π_1,V,K_1) is also a Stinespring representation. It enjoys one additional property, namely, that K_1 is the closed linear span of $\pi_1(A)VH$. Whenever the space of the representation enjoys this additional property, we call the triple a <u>minimal</u>, <u>Stinespring representation</u>. The following result summarizes the importance of this minimality condition.

<u>Proposition 4.2</u>. Let $\phi\colon A \to L(H)$ be completely positive and let (π_i,V_i,K_i) , $i = 1, 2$, be two minimal, Stinespring representations. Then there exists a unitary $U\colon K_1 \to K_2$ satisfying $UV_1 = V_2$ and $U\pi_1 U^* = \pi_2$.

<u>Proof</u>. If U exists, then necessarily,

$$U(\Sigma_i\ \pi_1(a_i)V_1 h_i) = \Sigma_i\ \pi_2(a_i)V_2 h_i .$$

Thus, it will be sufficient to verify that the above formula yields a well-defined isometry from K_1 to K_2 , since by the minimality condition, U will have dense range and hence be onto.

To this end, note that

46

$$\| \Sigma_i \ \pi_1(a_i)V_1 h_i \|^2 = \Sigma_{i,j} <V_1^* \pi_1(a_i^* a_j)V_1 h_j, \ h_i>$$

$$= \Sigma_{i,j} <\phi(a_i^* a_j)h_j, \ h_i> = \| \Sigma_i \ \pi_2(a_i)V_2 h_i \|^2 \ ,$$

so U is isometric and consequently well-defined, which is all that we needed to show.   □

We now show how some other dilation theorems can be deduced from Stinespring's result.

Theorem 4.3 (Sz.-Nagy). Let $T \in L(H)$ with $\|T\| \leq 1$. Then there exists a Hilbert space K containing H as a subspace and a unitary U on K with the property that K is the smallest closed reducing subspace for U containing H such that

$$T^n = P_H U^n|_H \ , \quad \text{for all non-negative integers } n \ .$$

Moreover, if (U',K') is another pair with the above properties, then there is a unitary $V: K \rightarrow K'$ such that $Vh = h$ for $h \in H$ and $VUV^* = U'$.

Proof. By Theorem 2.6, the map $\phi(p + \bar{q}) = p(T) + q(T)^*$, where p and q are polynomials, extends to a positive map of $C(\mathbb{T})$ into $L(H)$. This map is completely positive by Theorem 3.10.

Let (π,V,K) be a minimal, Stinespring representation of ϕ and recall that since $\phi(1) = 1$, we may identify VH with H . Setting $\pi(z) = U$, where z is the coordinate function, we have that U is unitary and that

$$T^n = \phi(z^n) = P_H \pi(z^n)|_H = P_H U^n|_H \ .$$

47

The minimality condition on (π,V,K) is equivalent to requiring that the span of

$$\{U^n H: n = 0, \pm1, \pm2, \ldots\}$$

be dense in K, which is equivalent to the requirement that there be no closed reducing subspace for U containing H other than K itself.

The final statement of the theorem is a consequence of the uniqueness of a minimal, Stinespring representation up to unitary equivalence, Proposition 4.2.   □

The techniques used to prove Theorem 4.3 can be used to prove a far more general result. Let $X \subseteq \mathbb{C}$ be a compact set and let $R(X)$ be the algebra of rational functions on X. An operator T is said to have a normal ∂X-dilation if there is a Hilbert space K containing H as a subspace and a normal operator N on K with $\sigma(N) \subseteq \partial X$ such that

$$r(T) = P_H r(N)|_H ,$$

for all r in $R(X)$. We shall call N a minimal normal ∂X-dilation for T, provided that K is the smallest reducing subspace for N that contains H. Clearly,

$$\|r(T)\| \leq \|r(N)\| \leq \sup \{|r(z)|: z \in \partial X\} ,$$

and so a necessary condition for T to have a normal ∂X-dilation is that X be a spectral set for T. It is a long-standing problem to determine if this condition is also sufficient [114]. It has only recently been verified that this condition is sufficient for X as simple a region as an annulus [1].

48

If $S = R(X) + \overline{R(X)}$ is dense in $C(\partial X)$, then $R(X)$ is called a Dirichlet algebra on ∂X. For example, if $\mathbb{C}/X$ has only finitely many components and the interior of X is simply-connected, then $R(X)$ is a Dirichlet algebra on ∂X. See [30] for this and further topological conditions on X which insure that $R(X)$ is a Dirichlet algebra on ∂X.

Theorem 4.4 (Berger-Foias-Lebow). Let $R(X)$ be a Dirichlet algebra on ∂X. If X is a spectral set for T, then T has a minimal, normal, ∂X-dilation. Moreover, any two minimal, normal, ∂X dilations for T are unitarily equivalent, via a unitary which leaves H invariant.

Proof. Let $\rho\colon R(X) \to L(H)$ be the unital contraction defined by $\rho(r) = r(T)$ so that $\tilde{\rho}\colon S \to L(H)$ is positive, where $S = R(X) + \overline{R(X)}$. Since S is dense in $C(\partial X)$ and positive maps are bounded, $\tilde{\rho}$ extends to a positive map ϕ on $C(\partial X)$. But by Theorem 3.10, ϕ is completely positive. The remainder of the proof precedes as in Theorem 4.3. □

To state the next dilation theorem, we need to introduce some notation. Let X be a compact Hausdorff space and let B be the σ-algebra of Borel sets on X. An $L(H)$-valued measure on X is a map $E\colon B \to L(H)$ which is weakly countably additive, that is, if $\{B_i\}$ is a countable collection of disjoint Borel sets with union B, then

$$\langle E(B)x,\ y\rangle = \Sigma_i\ \langle E(B_i)x,\ y\rangle,$$

for all x, y in H. The measure is bounded provided that

$$\sup\ \{\|E(B)\|\colon B \in B\} < \infty,$$

and we let $\|E\|$ denote this supremum. The measure is <u>regular</u>, provided that for all x , y in H , the complex measure given by

$$(*) \quad \mu_{x,y}(B) = <E(B)x, y>$$

is regular.

Given a regular, bounded, $L(H)$-valued measure E , one obtains a bounded, linear map

$$\phi_E: C(X) \to L(H)$$

via

$$(**) \quad <\phi_E(f)x, y> = \int fd\mu_{x,y} .$$

Conversely, given a bounded, linear map $\phi: C(X) \to L(H)$, if one defines regular Borel measures $\{\mu_{x,y}\}$ for each x and y in H by the above formula $(**)$, then for each Borel set B , there exists a unique, bounded operator $E(B)$, defined by formula $(*)$, and the map $B \to E(B)$ defines a bounded, regular $L(H)$-valued measure. Thus, we see that there is a one-to-one correspondence between the bounded, linear maps of $C(X)$ into $L(H)$ and the regular bounded $L(H)$-valued measures. Such measures are called;

    i)   <u>spectral</u>, if $E(M \cap N) = E(M) \cdot E(N)$,

    ii) <u>positive</u>, if $E(M) \geq 0$,

    iii) <u>self-adjoint</u>, if $E(M)^* = E(M)$,

for all Borel sets M and N .

Note that if E is spectral and self-adjoint, then $E(M)$ must be an orthogonal projection.

50

The following proposition, whose proof we leave to the reader (Exercise 4.10), summarizes the relationships between the above properties of measures and the properties of their corresponding linear maps.

Proposition 4.5. Let E be a regular bounded $L(H)$-valued measure and let $\phi: C(X) \to L(H)$ be its' corresponding linear map, then:

    i) ϕ is a homomorphism if and only if E is spectral.

    ii) ϕ is positive if and only if E is positive.

    iii) ϕ is self-adjoint if and only if E is self-adjoint.

    iv) ϕ is a *-homomorphism if and only if E is self-adjoint and spectral.

The correspondence between these measures and linear maps leads to a dilation result for these measures.

Theorem 4.6 (Naimark). Let E be a regular, positive, $L(H)$-valued measure on X . Then there exists a Hilbert space K , a bounded linear operator $V: H \to K$, and a regular, self-adjoint, spectral, $L(K)$-valued measure on X , F , such that

$$E(B) = V^* F(B)V .$$

Proof. Let $\phi: C(X) \to L(H)$ be the positive, linear map corresponding to E . Then ϕ is completely positive by Theorem 3.10, and so we may apply Stinespring's theorem to obtain a *-homomorphism $\pi: C(X) \to L(K)$ and a bounded, linear operator $V: H \to K$, such that $\phi(f) = V^* \pi(f)V$ for all f in $C(X)$. If we let F be the $L(K)$-valued measure corresponding to

π , then it is easy to verify that F has the desired properties. $\square$

As another application of Stinespring's theorem, we shall give a characterization of the completely positive maps between two matrix algebras. In contrast, there is an entire pantheon of classes of positive maps between matrix algebras and much that is not known about the relationships between these various classes (see, for example [20] and [125]).

We begin by remarking that if $\pi: M_n \to L(K)$ is a *-homomorphism, then up to unitary equivalence, K decomposes as an orthogonal, direct sum of n-dimensional subspaces such that π is the identity representation on each of the subspaces (Exercise 4.11).

Now let $\phi: M_n \to M_k = L(\mathbb{C}^k)$ and let (π, V, K) be a minimal, Stinespring representation for ϕ . By the construction of the space K given in Theorem 4.1, we see that $\dim (K) \le \dim (M_n \otimes \mathbb{C}^k) = n^2 k$.

Thus, up to unitary equivalence, we may decompose K into the direct sum of fewer than nk subspaces, each of dimension n , such that $\pi: M_n \to L(K)$ is the identity representation on each one. So let us write $K = \Sigma_{i=1}^{\ell} \oplus \mathbb{C}_i^n$, $\ell \le nk$, and let P_i denote the projection of K onto $\mathbb{C}_i^n$. We have that for any A in M_n ,

$$P_i \pi(A) |_{\mathbb{C}_i^n} = A .$$

Now, if we let $V_i: \mathbb{C}^k \to \mathbb{C}_i^n$ be defined by $V_i = P_i V$, then

$$\phi(A) = V^* \pi(A) V = \Sigma_{i,j=1}^{\ell} V_i^* \pi(A) V_j = \Sigma_{i=1}^{\ell} V_i^* A V_i ,$$

after identifying each $\mathbb{C}_i^n$ with $\mathbb{C}^n$.

We summarize these calculations in the following:

52

Proposition 4.7 (Choi). Let $\phi: M_n \to M_k$ be completely positive. Then there exists fewer than nk linear maps, $V_i: \mathbb{C}^k \to \mathbb{C}^n$, such that $\phi(A) = \Sigma_i V_i^* A V_i$ for all A in M_n.

There is another dilation theorem due to Naimark whose proof is closely related to the proof of Stinespring's theorem. Let G be a group and let $\phi: G \to L(H)$. We call ϕ completely positive definite if for every finite set of elements of G, $g_1, \ldots, g_n$, the operator matrix $(\phi(g_i^{-1}g_j))$ is positive.

Theorem 4.8 (Naimark). Let G be a topological group and let $\phi: G \to L(H)$ be weakly continuous and completely positive definite. Then there exists a Hilbert space K, a bounded operator $V: H \to K$, and a weakly continuous unitary representation $\rho: G \to L(H)$ such that

$$\phi(g) = V^* \rho(g) V .$$

Proof. Consider the vector space $C_0(G,H)$ of finitely supported functions from G to H and define a bilinear function on this space by

$$\langle f_1, f_2 \rangle = \Sigma_{g,g'} \langle \phi(g^{-1}g')f_1(g'), f_2(g) \rangle_H .$$

As in the proof of Stinespring's theorem, we have that $\langle f,f \rangle \geq 0$ and that the set $N = \{f \mid \langle f,f \rangle = 0\}$ is a subspace of $C_0(G,H)$. We let K be the completion of $C_0(G,H)/N$ with respect to the induced inner product.

For h in H, define Vh by

$$(Vh)(g) = \begin{cases} h, & \text{if } g = e \\ 0, & \text{if } g \neq e \end{cases} ,$$

53

where e denotes the identy of G , and let $\rho: G \to L(K)$ be left translation, that is,

$$(\rho(g)f)(g') = f(gg') .$$

It is straightforward to check that V is bounded and linear, that ρ is a unitary representation, and that $\phi(g) = V^*\rho(g)V$.

It remains to show that ρ is weakly continuous. Let $\{g_\lambda\}$ be a net in G which converges to g_0 . Since ρ is a unitary representation, it will suffice to show that $\rho(g_\lambda)$ converges weakly to $\rho(g_0)$ on a dense subspace.

Let f_1 , f_2 be in $C_0(G,H)$, then,

$$<\rho(g_\lambda)f_1, f_2> = \Sigma_{g,g'} \; <\phi(g^{-1}g')f_1(g_\lambda g'), f_2(g)>_H$$

$$= \Sigma_{g,g''} \; <\phi(g^{-1}g_\lambda^{-1}g'')f_1(g'), f_2(g)>_H$$

which, since all the sums involved in the above expressions are finite, converges to

$$\Sigma_{g,g''} \; <\phi(g^{-1}g_0^{-1}g'')f_1(g''), f_2(g)>_H = <\rho(g_0)f_1, f_2> .$$

Thus, we see that $\rho(g_\lambda)$ converges weakly to $\rho(g_0)$. This completes the proof of the theorem. $\square$

As with Stinespring's representation, there is a minimality condition that guarantees the uniqueness of this representation up to unitary equivalence (Exercise 4.12).

A map $\phi: G \to L(H)$ will be called positive definite if for every choice of n elements of G , $g_1, \ldots, g_n$, and scalars, $\alpha_1, \ldots, \alpha_n$, the operator

$$\Sigma_{i,j} \ \bar{\alpha}_i \alpha_j \phi(g_i^{-1}g_j)$$

is positive. We remark that this is equivalent to requiring that for every x in H, the map $\phi_x: G \to \mathbb{C}$, given by $\phi_x(g) = \langle\phi(g)x, x\rangle$, is completely positive definite.

We caution the reader that our terminology is not standard. What we have chosen to call completely positive definite is usually called simply positive definite, and the concept that we have introduced above and called positive definite is usually not introduced at all. Our rationale for this slight deviation in notation will be clear in 4.11, where we will show a correspondence between the two classes of maps on G that are defined above and maps on a certain C^*-algebra associated with G, $C^*(G)$. Not surprisingly, this correspondence will carry (completely) positive definite maps on G to (completely) positive maps on $C^*(G)$.

We begin by describing this correspondence in one case of particular interest. Let Z^n be the Cartesian product of n copies of the integers and let Π^n be the Cartesian product of n copies of the circle. Let $J = (j_1, \ldots, j_n)$ be in Z^n and let λ_j be the j-th coordinate function on Π^n. We set $\Lambda^J = \lambda_1^{j_1} \ldots \lambda_n^{j_n}$. Note that there is a one-to-one correspondence between unitary representations $\rho: Z^n \to L(H)$ and *-homomorphisms $\pi: C(\Pi^n) \to L(H)$, given by $\pi(\lambda_j) = \rho(e_j)$, where e_j is the n-tuple which is 1 in the j-th entry and 0 in the remaining entries.

Proposition 4.9. Let $\phi: Z^n \to L(H)$ be (completely) positive definite. Then there is a uniquely defined, (completely) positive map $\psi: C(\Pi^n) \to L(H)$, given by $\psi(\Lambda^J) = \phi(J)$. Conversely, if the (completely) positive map ψ is given, then the above formula defines a (completely) positive definite function ϕ on Z^n .

Proof. First, we consider the case where ϕ is completely positive definite. Let (ρ, V, K) be the Naimark dilation of ϕ , so that $\phi(J) = V^*\rho(J)V$, and let $\pi: C(\Pi^n) \to L(H)$ be the *-homomorphism associated with ρ . If we set $\psi(f) = V^*\pi(f)V$, then we obtain a map $\psi: C(\Pi^n) \to L(H)$, which is completely positive. Moreover, ψ satisfies

$$\psi(\Lambda^J) = V^*\pi(\Lambda^J)V = V^*\rho(J)V = \phi(J) .$$

The proof of the converse in the completely positive case is identical.

Now, suppose that ϕ is only positive definite. If we fix x in H and set $\phi_x(J) = <\phi(J)x, x>$, then ϕ_x is a completely positive definite function on Z^n . Thus, by the above there is a positive map $\psi_x: C(\Pi^n) \to \mathbb{C}$, with $\psi_x(\Lambda^J) = \phi_x(J)$. For fixed f in $C(\Pi^n)$, the set $\{\psi_x(f)\}$ as x varies over H forms a bounded, quadratic family and hence there exists a bounded operator $\psi(f)$ such that $<\psi(f)x, x> = \psi_x(f)$. This defines a linear map $\psi: C(\Pi^n) \to L(H)$ which is easily seen to be positive.

The converse in the positive case is similar. □

Corollary 4.10. For Z^n , the sets of positive definite and completely positive definite, operator-valued functions coincide.

4.11 Group C^*-algebras. The above results are part of a more general duality. Let G be a locally compact group and let dg be a (left) Haar measure on G . The space of integrable functions, $L^1(G)$, on G can be made into a *-algebra by defining

$$f_1 \times f_2(g') = \int f_1(g)f_2(g^{-1}g') \, dg \,,$$

and a *-operation by

$$f^*(g) = \Delta(g)^{-1}\overline{f(g^{-1})} \,,$$

where $\Delta(\cdot)$ is the modular function. It is then possible to endow $L^1(G)$ with a norm such that the completion of $L^1(G)$ is a C^*-algebra, denoted by $C^*(G)$ (See [34] or [89]).

There is a one-to-one correspondence between weakly continuous, unitary representations, $\rho: G \to L(H)$, and *-homomorphisms, $\pi: C^*(G) \to L(H)$, given by

$$\pi(f) = \int f(g)\rho(g) \, dg$$

when f is in $L^1(G)$. See [89] for a development of this theory.

In a similar fashion, there are one-to-one correspondences between the weakly continuous, (completely) positive definite, operator-valued functions on G and the (completely) positive maps on $C^*(G)$, given by

$$\psi(f) = \int f(g)\phi(g) \, dg$$

for f in $L^1(G)$. The proof that the above formula defines a one-to-one correspondence between these two classes of maps is similar to the proof of Proposition 4.9 and is left as an exercise (Exercise 4.15).

Proposition 4.9 follows from the above correspondence and the fact [89] that $C^*(Z^n) = C(\Pi^n)$.

Naimark's two dilation theorems ([80] and [81]) preceded Stinespring's dilation theorem [104]. Stinespring [104] defined completely positive maps on C^*-algebras, proved that positive maps on $C(X)$ were completely positive (Theorem 3.10) and then proved Theorem 4.1 as a generalization of Naimark's dilation theorem for positive, operator-valued measures.

Arveson [3] realized the important role that the theory of completely positive maps can play in the study of normal ∂X-dilations and gave the proofs of Theorems 4.3 and 4.4 that are presented here.

Other proofs of Sz.-Nagy's dilation theorem have used the theory of positive, definite functions on Z or "geometric" techniques, where the unitary and the space that it acts on are explicitly constructed ([111] and [97]). Two beautiful results of the geometric dilation techniques are Ando's dilation theorem for commuting pairs of contractions [2] and the Sz.-Nagy-Foias commutant lifting theorem [114].

The commutant lifting theorem says that if T is a contraction on a Hilbert space H, U is its minimal, unitary dilation, and X is an operator that commutes with T, then there exists an operator Y defined on the space on which U acts, commuting with U, such that $\|X\| = \|Y\|$ and

$$X = P_H Y |_H \, .$$

Ando's dilation theorem says that if T_1 and T_2 are commuting contractions on a Hilbert space H, then there exist commuting unitaries U_1 and U_2 such that

$$T_1^n T_2^m = P_H U_1^n U_2^m |_H$$

58

for all non-negative integers n and m . The analogue of this result

for 3 or more commuting contractions has been shown to be false by

Parrott [83] (see also, 6.9).

Note that Ando's result implies that the analogue of von Neumann's

inequality holds for commuting pairs of contractions, that is,

$$\|p(T_1,T_2)\| \leq \sup \{|p(z_1,z_2)| : |z_i| \leq 1 , \quad i = 1, 2\}$$

for all polynomials in 2 variables. The analogue of von Neumann's

inequality for 3 or more commuting contractions has been shown to be

false by Crabbe-Davies [31] and by Varopolos [119].

The correspondence between bounded, regular, operator-valued measures

on a compact Hausdorff space X and bounded, linear maps on C(X) is

discussed in Hadwin [47].

Proposition 4.7 was proved by Choi [18], who also developed the

theory of multiplicative domains [17] (Exercise 4.2).

See Bunce [15], for more on Korovkin type theorems (Exercise 4.9).

Exercise 4.16 is from Berger [9].

EXERCISES

4.1 Use Stinespring's representation theorem to prove that $\|V\|^2 = \|\phi\|_{cb}$

when ϕ is completely positive. Also, use the representation theorem

to prove that $\phi(a)^*\phi(a) \leq \|\phi(1)\|^2\phi(a^*a)$.

4.2 (Multiplicative Domains) Let A be a C^*-algebra with unit and let

$\phi: A \rightarrow L(H)$ be completely positive, $\|\phi(1)\| = 1$, with minimal

Stinespring representation, (π,V,K) .

(i) Prove that $\phi(a)^*\phi(a) = \phi(a^*a)$ if and only if VH is an invariant subspace for $\pi(a)$.

(ii) Show that $\{a \in A: \phi(a)^*\phi(a) = \phi(a^*a)\} =$
$\{a \in A: \phi(ba) = \phi(b)\phi(a)$ for all $b \in A\}$, and that this set is an algebra.

(iii) Show that $\{a \in A: \phi(a)^*\phi(a) = \phi(a^*a)$ and $\phi(a)\phi(a)^* = \phi(aa^*)\}$
$= \{a \in A: \phi(ba) = \phi(b)\phi(a)$ and $\phi(ab) = \phi(a)\phi(b)$ for all
$b \in A\}$ is a C^*-subalgebra of A . This subalgebra is called the <u>multiplicative domain of</u> ϕ .

4.3 (Bimodule Maps) Let A , B and C be C^*-algebras with unit and suppose that C is contained in both A and B , with $1_C = 1_A$
and $1_C = 1_B$. A linear map $\phi: A \to B$ is called a <u>C-bimodule</u> map
if $\phi(c_1 a c_2) = c_1\phi(a)c_2$ for all c_1 , c_2 in C . Let $\phi: A \to B$
be completely positive.

(i) If $\phi(1) = 1$, then prove that ϕ is a C-bimodule map if and only if $\phi(c) = c$ for all c in C .

(ii) Prove, in general, that ϕ is a C-bimodule map if and only if $\phi(c) = c \cdot \phi(1)$ for all c in C .

4.4 Let C be the C^*-subalgebra of diagonal matrices in M_n . Prove that a linear map $\phi: M_n \to M_n$ is a C-bimodule map if and only if ϕ
is the Schur product map, S_T , for some matrix T .

4.5 Let $\phi: A \to A$ be completely positive with $\phi(1) = 1$.

(i) Show that if $\phi(a) = a$, then a is in the multiplicative domain.

(ii) Show that $B = \{a: \phi(a) = a\}$ is a C^*-subalgebra, but that in general, B is distinct from the multiplicative domain of ϕ .

(iii) Show that ϕ is a B-bimodule map.

60

4.6 Let $\phi: G \to L(H)$ be completely positive definite. Prove that ϕ is weakly continuous if and only if ϕ is strongly continuous.

4.7 Let $\phi: G \to M_n$ be continuous. Prove that ϕ is completely positive definite if and only if there exists a Hilbert space H and continuous functions $x_i: G \to H$, $i = 1, \ldots, n$, such that
$$\phi(g^{-1}g') = (<x_j(g'), x_i(g)>) \; .$$

4.8 A semigroup G with an involution is called a *-semigroup. A function $\phi: G \to L(H)$ is called <u>completely</u> <u>positive</u> <u>definite</u> if $(\phi(g_i^* g_j))$ is positive for every set of finitely many elements, $g_1, \ldots, g_n$, of G and <u>bounded</u> if $(\phi(g_i^* g^* g g_j)) \leq M_g(\phi(g_i^* g_j))$ where M_g is a constant depending only on g . Prove a version of Naimark's dilation theorem for *-semigroups.

4.9 Let A be a C^* -algebra with unit and let $\phi_n: A \to L(H)$ be a sequence of completely positive maps such that $\phi_n(1) \to 1$ in the weak operator topology.

  (i) Prove that $\{a: \phi_n(a)^* \phi_n(a) - \phi_n(a^* a) \to 0$ and $\phi_n(a)\phi_n(a)^* - \phi_n(aa^*) \to 0$ in the weak operator topology$\}$ is a C^* -subalgebra of A .

  (ii) Let $\phi_n: C([0,1]) \to \mathbb{C}$ be a sequence of positive, linear functionals. Prove that if $\phi_n(1) \to 1$, $\phi_n(t) \to 1/2$, and $\phi_n(t^2) \to 1/3$, then $\phi_n(f) \to \int_0^1 f(t) \, dt$, for all f in C([0,1]).

  (iii) (Korovkin) Prove that if $\phi_n: C([0,1]) \to C([0,1])$ is a sequence of positive maps such that $\phi_n(1)$, $\phi_n(t)$, and $\phi_n(t^2)$ converges in norm to 1 . t , and t^2 , respectively, then $\phi_n(f)$ converges in norm to f , for all f .

4.10 Prove Proposition 4.5.

4.11 Let $\pi: M_n \to L(K)$ be a unital *-homomorphism. Prove that up to unitary equivalence, $K = H_1 \oplus \ldots \oplus H_n$, with $H_i = H$, $i = 1, \ldots, n$, such that $\pi(E_{i,j})$ is the identity map from H_j to H_i. Show that up to unitary equivalence, π is the direct sum of $\dim(H)$ copies of the identity map.

4.12 Give a minimality condition for the Naimark representation of completely positive, definite functions on a group and prove uniqueness of minimal representations up to unitary equivalence.

4.13 Let $t \to A(t)$, $t \geq 0$, be a weakly continuous semi-group of contraction operators, $A(0) = I$. For $t < 0$, set $A(t) = A(-t)^*$. Prove that this extended map is completely positive definite on $\mathbb{R}$. What does Naimark's dilation theorem imply? [Hint: Recall the proof of Theorem 2.6.]

4.14 (Trigonometric Moments) Let $\{a_n\}_{n=-\infty}^{+\infty}$ be a sequence of complex numbers. Prove that there exists a positive measure μ on Π such that $a_n = \int_\Pi z^n d\mu(z)$ if and only if $a(n) = a_n$ is a positive, definite function on Z.

4.15 Verify the claims of 4.11.

4.16 (Berger) Let T be an operator on a Hilbert space H. Prove that $w(T) \leq 1$ if and only if there exists a Hilbert space K containing H and a unitary operator U on K such that $T^n = 2P_H U^n|_H$ for all positive integers.

5 Completely positive maps into M_n

In this chapter we characterize completely positive maps into M_n. This characterization allows us to prove an extension theorem for completely positive maps and lends further insight into the properties of positive maps on operator systems. These results are all a consequence of a duality between maps into M_n and linear functionals.

Let M be an operator space and let $\{e_j\}_{j=1}^n$ be the canonical basis for $\mathbb{C}^n$. If A is in M_n, then let $A_{(i,j)}$ denote the (i,j)-entry of A so that $A_{(i,j)} = <Ae_j,e_i>$. If $\phi: M \to M_n$ is a linear map, then we associate a linear functional s_ϕ on $M_n(M)$ to ϕ by the following formula,

$$s_\phi((a_{i,j})) = 1/n \ \Sigma_{i,j} \ \phi(a_{i,j})_{(i,j)} \ .$$

Alternatively, if we let x denote the vector in $\mathbb{C}^{n^2}$ given by $x = e_1 \oplus \ldots \oplus e_n$, then

$$s_\phi((a_{i,j})) = 1/n<\phi_n((a_{i,j}))x, \ x> \ ,$$

where the inner product is taken in $\mathbb{C}^{n^2}$.

We leave it to the reader to verify that $\phi \to s_\phi$ defines a linear map from $L(M,M_n)$, the vector space of linear maps from M into M_n, into the vector space $L(M_n(M),\mathbb{C})$. If M contains the unit and $\phi(1) = 1$, then $s_\phi(1) = 1$.

Finally, if $s: M_n(M) \to \mathbb{C}$, then we define $\phi_s: M \to M_n$ via

$$(\phi_s(a))_{(i,j)} = n \cdot s \ (a \otimes E_{i,j}) \ ,$$

where $a \otimes E_{i,j}$ is the element of $M_n(M)$ which has a for its (i,j)-entry and is 0 elsewhere.

We leave it to the reader to verify that these two operations are mutual inverses.

Theorem 5.1. Let A be a C^*-algebra with unit, let S be an operator system in A, and let $\phi: S \to M_n$. The following are equivalent:

    i) ϕ is completely positive.

    ii) ϕ is n-positive.

    iii) s_ϕ is positive.

Proof. Obviously, i) implies ii). Also, ii) implies iii) is clear by the alternate definition of s_ϕ. So assume that s_ϕ is positive and we shall prove that ϕ is completely positive.

By Krein's theorem (Exercise 2.10), we may extend s_ϕ from $M_n(S)$ to a positive, linear functional s on $M_n(A)$. Since s extends s_ϕ, the map $\psi: A \to M_n$ associated with s extends ϕ. Clearly, if we can prove that ψ is completely positive, then ϕ will be completely positive.

To verify that ψ is m-positive by Lemma 3.11, it is sufficient to consider a positive element of $M_m(A)$ of the form $(a_i^* a_j)$. Since $\psi_m((a_i^* a_j))$ acts on $\mathbb{C}^{mn}$, to see that it is positive, it is sufficient to take $x = x_1 \oplus \ldots \oplus x_m$, where each $x_j = \Sigma_k \lambda_{j,k} e_k$ is in $\mathbb{C}^n$ and calculate

$$\langle \psi_m((a_i^* a_j))x, x \rangle = \Sigma_{i,j} \langle \psi(a_i^* a_j)x_j, x_i \rangle$$

(*)
$$= \Sigma_{i,j,k,\ell} \lambda_{j,k} \bar{\lambda}_{i,\ell} \langle \psi(a_i^* a_j)e_k, e_\ell \rangle$$

$$= \Sigma_{i,j,k,\ell} \lambda_{j,k} \bar{\lambda}_{i,\ell} s(a_i^* a_j \otimes E_{\ell,k}) .$$

64

Let A_i be the $n \times n$ matrix whose first row is $(\lambda_{i,1}, \ldots, \lambda_{i,n})$ and whose remaining rows are 0 . We have that

$$A_i^* A_j = \Sigma_{k,\ell} \, \bar{\lambda}_{i,\ell} \lambda_{j,k} E_{\ell,k} \; ,$$

and thus (*) becomes

$$\Sigma_{i,j} \; s(a_i^* a_j \otimes A_i^* A_j) = s((\Sigma_i \, a_i \otimes A_i)^* (\Sigma_j \, a_j \otimes A_j)) \geq 0 \; ,$$

since s is positive. Thus, ψ is m-positive for all m .   □

Theorem 5.2. Let A be a C^*-algebra with unit, S an operator system contained in A , and $\phi: S \to M_n$ completely positive. Then there exists a completely positive map $\psi: A \to M_n$ extending ϕ .

Proof. Let s_ϕ be the positive, linear functional on $M_n(S)$ associated with ϕ and extend it to a positive, linear functional s on $M_n(A)$ by Krein's theorem. By Theorem 5.1, the map ψ associated with s is completely positive. Finally, since s extends s_ϕ , it is easy to see that ψ extends ϕ .   □

In fact, the proof of 5.1 contains 5.2.

We now restate these results for operator spaces.

Theorem 5.3. Let A be a C^*-algebra with unit, let M be a subspace of A containing 1 , and let $\phi: M \to M_n$ with $\phi(1) = 1$. The following are equivalent:

i) ϕ is completely contractive.

ii) ϕ is n-contractive.

iii) s_ϕ is contractive.

Proof.   Let $S = M + M^*$.   Clearly,   i)   implies   ii)   and   ii)
implies   iii) .   Assuming   iii) ,   since s_ϕ is unital and contractive,
we may extend it to a positive, unital map $\tilde{s}_\phi$ on $M_n(M) + M_n(M)^* = M_n(S)$.
By Theorem 5.1, the linear functional $\tilde{s}_\phi$ is associated with a completely
positive map on S .   This map is readily seen to be $\tilde{\phi}$.   Hence, $\tilde{\phi}$ is
completely positive and so ϕ must be completely contractive.   □

Theorem 5.4.   Let A , M , and ϕ satisfy the conditions of
Theorem 5.4. Then ϕ extends to a completely positive map on A .

Proof.   In the proof of Theorem 5.3, we saw that $\tilde{\phi}$ is completely
positive and hence extends to a completely positive map on A by
Theorem 5.2.   □

There is one way in which the above correspondence between linear
functionals on $M_n(S)$ and linear maps of S into M_n is not well-behaved.
Suppose that $s: M_n(S) \to \mathbb{C}$ is positive and unital, so that $\|s\| = 1$.
Then s gives rise to a completely positive map $\phi_s: S \to M_n$, but ϕ_s is
not necessarily unital.   Indeed, since $\phi_s(1)_{(i.j)} = ns(1 \otimes E_{i,j}) = ns(E_{i,j})$,
we have that ϕ_s is unital if and only if

$$s(E_{i,j}) = \begin{cases} 1/n , & i = j \\ 0 , & i \neq j \end{cases} .$$

Because of this fact, ϕ_s is not necessarily a contraction. In fact, it is not hard to construct examples of a unital, positive s such that $\|\phi_s\| = n$ (Exercise 5.1).

To obtain generalizations of the above results to the case where M_n is replaced by $L(H)$ requires some topological preliminaries which we postpone until the next chapter. We turn instead to some other applications of this correspondence between linear functionals and linear maps.

Let S be an operator system and S^+ its cone of positive elements. We define

$$S^+ \otimes M_n^+ = \{ \Sigma_i\, p_i \otimes Q_i : p_i \in S^+ , Q_i \in M_n^+ \} ,$$

where the above sum is finite. Note that $S^+ \otimes M_n^+$ is a cone contained in the cone $M_n(S)^+$ of positive elements of $M_n(S)$.

We have seen above that if $\phi : S \to M_n$, then ϕ is completely positive if and only if its associated linear functional s_ϕ is positive on $M_n(S)^+$. The following points out the relevance of the set defined above.

Lemma 5.5. Let $\phi : S \to M_n$. Then ϕ is positive if and only if $s_\phi : M_n(S) \to \mathbb{C}$ assumes positive values on $S^+ \otimes M_n^+$.

Proof. Let $\phi : S \to M_n$ be positive, let p be in S^+ , and let Q be in M_n^+ . In order to prove that s_ϕ assumes positive values on $S^+ \otimes M_n^+$, it will be sufficient to prove that $s_\phi(p \otimes Q)$ is positive. Furthermore, since by Lemma 3.11, Q can be written as a convex sum of matrices of the form $(\bar{\alpha}_i \alpha_j)$, it will suffice to assume that $Q = (\bar{\alpha}_i \alpha_j)$. Thus, $p \otimes Q = (\bar{\alpha}_i \alpha_j p)$ and

$$n \cdot s_\phi(p \otimes Q) = \Sigma_{i,j} \, \phi(\bar{\alpha}_i\alpha_jp)_{(i,j)} = \Sigma_{i,j} \, \bar{\alpha}_i\alpha_j<\phi(p)e_j,e_i> = <\phi(p)x,x> \geq 0 \, ,$$

where $x = \alpha_1e_1 + \ldots + \alpha_ne_n$.

Conversely, assume that s_ϕ is positive on $S^+ \otimes M_n^+$, let p be in S^+ , and let $x = \alpha_1e_1 + \ldots + \alpha_ne_n$ be a vector in $\mathbb{C}^n$. We have that

$$<\phi(p)x, \, x> = \Sigma_{i,j} \, <\phi(\bar{\alpha}_i\alpha_jp)e_j, \, e_i> = n \cdot s_\phi((\bar{\alpha}_i\alpha_jp)) \geq 0 \, .$$

since $(\bar{\alpha}_i\alpha_jp)$ is in $S^+ \otimes M_n^+$. Thus, ϕ is positive.   □

Theorem 5.6. Let S be an operator system. Then the following are equivalent:

   i) Every positive map $\phi: S \to M_n$ is completely positive.

   ii) Every unital, positive map $\phi: S \to M_n$ is completely positive.

   iii) $S^+ \otimes M_n^+$ is dense in $M_n(S)^+$.

Proof. Clearly, i) implies ii) . The proof that ii) implies i) is left as an exercise (Exercise 5.2) . We prove the equivalence of i) and iii) .

If $S^+ \otimes M_n^+$ is dense in $M_n(S)^+$ and $\phi: S \to M_n$ is positive, then, by Lemma 5.5, s_ϕ will be positive on $M_n(S)^+$, and so, by Theorem 5.1, ϕ is completely positive.

Conversely, if $S^+ \otimes M_n^+$ is not dense in $M_n(S)^+$, then fix p in $M_n(S)^+$, but not in the closure of $S^+ \otimes M_n^+$. By the Krein-Milman theorem, there will exist a linear functional s on $M_n(S)$, which is positive on $S^+ \otimes M_n^+$, but such that $s(p) < 0$. The linear map $\phi_s: S \to M_n$, induced by s , is then positive by Lemma 5.5, but not completely positive.   □

Corollary 5.7. Let S be an operator system. Then the following are equivalent:

    i) For every C^*-algebra B, every positive $\phi: S \to B$ is completely positive.

    ii) For all n, every positive $\phi: S \to M_n$ is completely positive.

    iii) $S^+ \otimes M_n^+$ is dense in $M_n(S)^+$ for all n.

Proof. Clearly, ii) and iii) are equivalent and i) implies ii). To see that ii) implies i), it is sufficient to consider $B = L(H)$. Given $(a_{i,j})$ in $M_n(S)^+$, to check that $\phi_n((a_{i,j}))$ is positive, it is enough to choose $x_1, \ldots, x_n$ in H and check that

$$\Sigma_{i,j} <\phi(a_{i,j})x_j, x_i> \geq 0 .$$

Let F be the finite dimensional subspace spanned by these n vectors, and let $\psi: S \to L(F) \cong M_k$, for some $k \leq n$, be the compression of ϕ to F. By ii), ψ is completely positive and hence,

$$0 \leq \Sigma_{i,j} <\psi(a_{i,j})x_j, x_i> = \Sigma_{i,j} <\phi(a_{i,j})x_j, x_i>$$

as desired. □

As well as being related to complete positivity, the above cone $S^+ \otimes M_n^+$ determines the norm behavior of positive maps. Let S be an operator system and let a be in S. We say that S has a partition of unity for a, and say that a is partitionable with respect to S, provided that for every $\epsilon > 0$, there exists positive elements, $p_1, \ldots, p_n$, in S, with $\Sigma_i p_i \leq 1$, and scalars. $\lambda_1, \ldots, \lambda_n$, with $|\lambda_i| \leq \|a\|$, such that

$$\| a - \Sigma_i \ \lambda_i p_i \| < \varepsilon \ .$$

We say that S has a __partition of unity__ for a subset M of S
provided that every element of M is partitionable with respect to S .

Lemma 5.8. Let S be an operator system, a in S . with $\|a\| \leq 1$.
Then the following are equivalent:

i) a is partitionable with respect to S .

ii) Every positive map ϕ with domain S satisfies $\|\phi(a)\| \leq \|\phi(1)\|$.

iii) $\begin{bmatrix} 1 & a \\ a^\star & 1 \end{bmatrix}$ is in the closure on $S^+ \otimes M_2^+$.

Proof. The proof that i) implies ii) is identical to the proof
of Theorem 2.4.

To see that ii) implies iii) , assume that iii) is not met. Let
$s : M_2(S) \to \mathbb{C}$ be a linear functional such that $s\left(\begin{bmatrix} 1 & a \\ a^\star & 1 \end{bmatrix} \right) < 0$, while
s is positive on $S^+ \otimes M_2^+$, and let $\phi : S \to M_2$ be the linear map
associated with s . By Lemma 5.5, ϕ is positive. Since

$$\left\langle \begin{bmatrix} \phi(1) & \phi(a) \\ \phi(a)^\star & \phi(1) \end{bmatrix} \begin{bmatrix} e_1 \\ e_2 \end{bmatrix} , \begin{bmatrix} e_1 \\ e_2 \end{bmatrix} \right\rangle = 2s\left(\begin{bmatrix} 1 & a \\ a^\star & 1 \end{bmatrix} \right) < 0 ,$$

we have that

$$\phi_2\left(\begin{bmatrix} 1 & a \\ a^\star & 1 \end{bmatrix} \right)$$

70

is not positive. If we construct a unital, positive map ϕ' from ϕ as prescribed by Exercise 5.2i), then by Exercise 5.2ii), we have that

$$\phi_2'\left(\begin{bmatrix} 1 & a \\ a^* & 1 \end{bmatrix}\right) = \begin{bmatrix} 1 & \phi'(a) \\ \phi'(a)^* & 1 \end{bmatrix}$$

is not positive. Thus, by Lemma 3.1, $\|\phi'(a)\| > 1 = \|\phi'(1)\|$, and so ii) does not hold.

Now suppose that iii) is true, and let $\varepsilon > 0$, P_i be in S^+ , Q_i be in M_2^+ , $i = 1, \ldots, n$, be given such that

$$\left\| \begin{bmatrix} 1 & a \\ a^* & 1 \end{bmatrix} - \Sigma_i \; P_i \otimes Q_i \right\| \; < \; \varepsilon \; . \qquad (*)$$

By perturbing the Q_i slightly, we may assume that $Q_i = \begin{bmatrix} r_i & \lambda_i \\ \bar{\lambda}_i & t_i \end{bmatrix}$,

with $r_i t_i \neq 0$ for all i . We set $s_i = (r_i + t_i)/2$. Note that if $s_i = 0$, then $\lambda_i = 0$. The equation $(*)$ implies the following inequalities;

$$1 - \varepsilon < \Sigma_i \; r_i P_i < 1 + \varepsilon \; ,$$

$$1 - \varepsilon < \Sigma_i \; t_i P_i < 1 + \varepsilon \; ,$$

$$\| a - \Sigma_i \; \lambda_i P_i \| < \varepsilon \; .$$

Thus,

$$1 - \varepsilon < \Sigma_i \; s_i P_i < 1 + \varepsilon \; ,$$

while

$$\| a - \Sigma_i \; (\lambda_i/s_i) s_i P_i \| < \varepsilon \; ,$$

71

with $|\lambda_i/s_i| \leq 1$, and we set $\lambda_i/s_i = 0$ when $s_i = 0$. These last two equations are clearly enough to guarantee that a is partitionable. □

Theorem 5.9. Let S be an operator system, M a subset of S . Then every positive map on S has norm $\|\phi(1)\|$ when restricted to M if and only if S has a partition of unity for M .

Proof. The proof is an immediate application of the above lemma. □

Corollary 5.10. Let B be a C^*-algebra with unit, let A be a subalgebra of B containing the unit, and let $S = A + A^*$. Then S has a partition of unity for A .

Proof. By Corollary 2.8, every positive map on S satisfies $\|\phi(a)\| \leq \|\phi(1)\| \cdot \|a\|$ for a in A . Thus, by Lemma 5.7, S has a partition of unity for A . □

We now study two examples to illustrate the use of the above ideas. The second arises from the theory of hypo-Dirichlet algebras, which we shall examine in more detail in Chapter 9.

Our first example is just a re-examination of an earlier example. Let S and ϕ be the operator system and positive map of Example 2.2. Since $\phi(1) = 1$, while $\|\phi(z)\| = 2$ with $\|z\| = 1$, we see that z is not partitionable with respect to S . It is interesting to attempt to show this directly.

The second example is somewhat longer.

Example 5.11. Let $0 < R_1 < R_2$ and let A denote the annulus in $\mathbb{C}$ with inner radius R_1 and outer radius R_2, i.e.,

$$A = \{z \in \mathbb{C}: R_1 \leq |z| \leq R_2\} .$$

We let $R(A)$, as before, denote the algebra of rational functions on A and let S denote the closure of $R(A) + \overline{R(A)}$ in $C(\partial A)$. By a theorem of Walsh [121], S is a codimension 1 subspace of $C(\partial A)$. We let $\Gamma_j = \{z: |z| = R_j\}$, $j = 1, 2$, denote the two boundary circles of A, and define positive linear functionals s_j, $j = 1, 2$, on $C(\partial A)$ by

$$s_j(f) = \tfrac{1}{2\pi} \int_0^{2\pi} f(R_j e^{it}) \, dt .$$

If $f(z) = \sum_{k=-n}^{+n} a_k z^k$ is a finite Laurent polynomial, then $s_1(f) = s_2(f) = a_o$. Since the finite Laurent polynomials are dense in $R(A)$ and since s_1 and s_2 are self-adjoint functionals, we have that $s_1(f) = s_2(f)$ for all f in S. Since S is of codimension 1, we have that

$$S = \{ f \in C(\partial A): s_1(f) = s_2(f)\} .$$

We claim that S has no partition of unity for S. Before proving this claim, let us explore some of the consequences of the claim.

Set

$$p(z) = \begin{cases} 0 , & z \in \Gamma_1 \\ 1 , & z \in \Gamma_2 \end{cases} ,$$

so that $s_1(p) = 0 \neq 1 = s_2(p)$. From this, we see that $p \notin S$, and so the span of S and p is all of $C(\partial A)$.

We leave it to the reader to verify that the claim that S has no partition of unity for S yields a positive, unital map $\phi: S \to M_2$ with the following properties (Exercise 5.4):

i) ϕ is not contractive.

ii) ϕ is not completely positive.

iii) ϕ has no positive extension to $C(\partial A)$.

iv) If $L_1 = \{f \in S: f \le p\}$, $L_2 = \{f \in S: f \ge p\}$, then there is no matrix P in M_2 satisfying

$$\phi(f_1) \le P \le \phi(f_2)$$

for all f_1 in L_1 , f_2 in L_2 .

v) If $\psi = \phi|_{R(A)}$, then ψ is a unital contraction, but $\tilde{\psi}$ is not contractive.

vi) ψ has no contractive extension to S , and so none to $C(\partial A)$ either.

vii) ψ is not 2-contractive.

Part vi) , in particular, shows that a direct generalization of the Hahn-Banach extension theorem to operator-valued mappings fails even when the domain is commutative.

To verify the claim, it will suffice to construct an element of S that is not partitionable. Fix a small positive number $\delta > 0$ and set $j = 1 , 2$,

$$X_j = \{R_j e^{it}: \delta \le t \le \pi - \delta\} ,$$

$$Y_j = \{R_j e^{it}: \pi + \delta \le t \le 2\pi - \delta\} .$$

Define a continuous function f on ∂A by setting $f(X_1) = 1$,

74

$f(Y_1) = -1$, $f(X_2) = i$, $f(Y_2) = -i$, and extending f linearly on the remaining arcs, so that $\|f\| = 1$. It is easy to check that $s_1(f) = s_2(f)$, so that f is in S .

Suppose that we were given scalars $\lambda_1, \ldots, \lambda_n$, $|\lambda_j| \leq 1$, $j = 1, \ldots, n$, and positive functions in S , $p_1, \ldots, p_n$, summing up to less than 1 such that f is not merely approximated by the sum, but such that

$$f = \Sigma_i \, \lambda_i p_i \ .$$

Since $f(X_1) = 1$ and $|\lambda_j| \leq 1$, we must have that some subset of the λ_j's is exactly equal to 1 , which after re-ordering, we may take to be $\lambda_1, \ldots, \lambda_k$. It is not difficult to see that the above equations imply that

$$p_1(x) + \ldots + p_k(x) = 1$$

for all x in X_1 . Similarly, after re-ordering the remaining λ_j's , we find a set, $\lambda_{k+1}, \ldots, \lambda_m$, which are all equal to -1 , and

$$p_{k+1}(x) + \ldots + p_m(x) = 1 \ ,$$

for all x in Y_1 . But this implies that

$$s_1(p_1 + \ldots + p_m) \geq 1 - 4\delta \ ,$$

and since s_1 and s_2 agree on S ,

$$s_2(p_{m+1} + \ldots + p_n) \leq 4\delta \ .$$

We have that $\mathrm{Im}(f) = \mathrm{Im}(\lambda_{m+1})p_{m+1} + \ldots + \mathrm{Im}(\lambda_n)p_n$, so

$$1 - 4\delta \leq s_2(|\mathrm{Im}(f)|) \leq s_2(p_{m+1} + \ldots + p_n) \leq 4\delta \ ,$$

75

a clear contradiction for $\delta < 1/8$.

Thus, we see that no sum of the above form could actually equal f . An analogous, but somewhat more detailed argument, shows that for the same δ and ε sufficiently small, no sum of the above form can approximate f to within ε . We leave it to the reader to verify this claim.

NOTES

The correspondence between linear maps from a space into M_n and linear functionals on the tensor product of that space with M_n , is a recurring theme in this subject (see for example, [20], [21], [68], [106]). Our presentation owes a great deal to [3] and [101].

The study of $S^+ \otimes M_n^+$ is adapted from [20], where it is used to study positive maps between matrix algebras. In contrast to Proposition 4.7, a number of unanswered questions and surprising examples arise in the study of positive maps between matrix algebras ([20], [105], and [125]) . For example, positive maps between M_n and M_k are characterized by linear functionals on $M_n(M_k)$, which are positive on $M_n^+ \otimes M_k^+$. A decent characterization of the matrices in $M_n(M_k) = M_{nk}$ that belong to this cone is still not available. See [20] and [126] for an introduction to this topic.

Example 5.11 is adapted from a similar example in [36]. The partition of unity techniques originate there also.

5.1 Let A be a unital C^*-algebra. Give an example of a linear functional,
     $s: M_n(A) \to \mathbb{C}$, such that s is unital and positive, but such that the
     associated linear map, $\phi_s: A \to M_n$, has norm n .

5.2 Let $\phi: S \to M_n$ be positive and set $\phi(1) = P$. Let Q be the
     projection onto the range of P and let R be positive with
     $(I - Q)R = 0$, $RPR = Q$.

     i)   Let $\psi: S \to M_n$ be any positive, unital map and set
          $\phi'(a) = R\phi(a)R + (I - Q)\psi(a)(I - Q)$, then ϕ' is a
          unital, positive map.

     ii) Show that if $(a_{i,j})$ is positive in $M_k(S)$, but $\phi_k((a_{i,j}))$ is
          not positive, then $\phi'_k((a_{i,j}))$ is not positive either.

     iii) Deduce the equivalence of i) and ii) in Theorem 5.6.

5.3 Assume that the equivalent conditions of Theorem 5.9 are not met.
     Show that then there always exists a unital, positive map
     $\phi: S \to M_2$ which is not contractive.

5.4 Verify the existence of a $\phi: S \to M_2$ with properties i) - vii) of
     Example 5.11.

5.5 Let S be an operator system. Prove that the following are equivalent:

     i)   For every C^*-algebra B , every positive $\phi: S \to B$ is n-positive.

     ii) $S^+ \otimes M_n^+$ is dense in $M_n(S)^+$.

5.6 Use Corollary 5.7 to give an alternate proof of the fact that every
     positive map with domain $C(X)$ is completely positive.

6 Arveson's extension theorems

In this chapter we extend the results of Chapter 5 from finite dimensional ranges, M_n , to maps with range $L(H)$. We then develop the immediate applications of the extension theorems to dilation theory. We begin with some observations of a general functional analytic nature.

Let X and Y be Banach spaces, let Y^* denote the dual of Y , and let $L(X,Y^*)$ denote the bounded linear transformations of X into Y^* . We wish to construct a Banach space such that $L(X,Y^*)$ is isometrically isomorphic to its dual. This will allow us to endow $L(X,Y^*)$ with a weak*-topology.

Fix vectors x in X and y in Y and define a linear functional $x \otimes y$ on $L(X,Y^*)$ by $x \otimes y (L) = L(x)(y)$. Since $|x \otimes y (L)| \leq \|L\| \cdot \|x\| \cdot \|y\|$, we see that $x \otimes y$ is in $L(X,Y^*)^*$ with $\|x \otimes y\| \leq \|x\| \|y\|$. In fact, $\|x \otimes y\| = \|x\| \|y\|$ (Exercise 6.1).

It is not difficult to check that the above definition is bilinear, i.e., $x \otimes (y_1 + y_2) = x \otimes y_1 + x \otimes y_2$, $(x_1 + x_2) \otimes y = x_1 \otimes y + x_2 \otimes y$. and $(\lambda x) \otimes y = x \otimes (\lambda y) = \lambda(x \otimes y)$ for $\lambda \in \mathbb{C}$. We let Z denote the closed linear span in $L(X,Y^*)^*$ of these elementary tensors. Actually, Z can be identified as the completion of $X \otimes Y$ with respect to a cross-norm (Exercise 6.1), but we shall not need that fact here.

Lemma 6.1, $L(X,Y^*)$ is isometrically isomorphic to Z^* with the duality given by:

$$<L, x \otimes y> = x \otimes y(L) .$$

Proof. It is straightforward to verify that the above pairing defines an isometric isomorphism of $L(X,Y^*)$ into Z^*. To see that it is onto, fix $f \in Z^*$, and for each x, define a linear map, $f_x: Y \to \mathbb{C}$, via $f_x(y) = f(x \otimes y)$. Since $|f_x(y)| \le \|f\| \|x\| \|y\|$, we have that $f_x \in Y^*$. It is straightforward to verify that if we set $L(x) = f_x$, then $L: X \to Y^*$ is linear, and that L is bounded with $\|L\| \le \|f\|$.

Thus, $L \in L(X,Y^*)$ and clearly under our correspondence $L \to f$, which completes the proof. □

We call the weak*-topology that is induced on $L(X,Y^*)$ by this identification the BW-topology (for bounded weak). The following Lemma explains the name.

Lemma 6.2. Let $\{L_\lambda\}$ be a bounded net in $L(X,Y^*)$. Then L_λ converges to L in the BW-topology if and only if $L_\lambda(x)$ converges weakly to $L(x)$ for all x in X.

Proof. If L_λ converges to L in the BW-topology, then $L_\lambda(x)(y) = \langle L_\lambda, (x \otimes y) \rangle \to \langle L, (x \otimes y) \rangle = L(x)(y)$ for all y in Y. Thus, $L_\lambda(x)$ converges weakly to $L(x)$ for all x.

Conversely, if $L_\lambda(x)$ converges weakly to $L(x)$ for all x, then $\langle L_\lambda, (x \otimes y) \rangle$ converges to $\langle L, (x \otimes y) \rangle$ for all x and y and hence on the linear span of the elementary tensors. But since the net is bounded, this implies that it converges on the closed linear span. □

If H is a Hilbert space, then $L(H)$ is the dual of a Banach space. This Banach space can be identified with the space of ultraweakly continuous, linear functionals or with the trace class operators (TC) on H with trace norm, $\|T\|_1 = \text{tr}(|T|)$, where $\text{tr}(\cdot)$ denotes the trace [30, Theorem I.4.5]. We prefer to focus on the duality with the trace class operators.

Under this duality, an operator $A \in L(H)$ is identified with the linear functional $\text{tr}(AT)$ for $T \in (TC)$. If h, k are in H, let $R_{h,k}$ denote the elementary rank one operator on H given by $R_{h,k}(x) = <x,k>h$. The linear span of these operators is dense in (TC) in the trace norm (Exercise 6.2) and for $A \in L(H)$,

$$\text{tr}(AR_{h,k}) = <Ah,k> .$$

Proposition 6.3. Let X be a Banach space and let H be a Hilbert space, then a bounded net $\{L_\lambda\}$ in $L(X,L(H))$ converges in the BW-topology to L if and only if $<L_\lambda(x)h,k>$ converges to $<L(x)h,k>$ for all h, k in H and x in X.

Proof. We have that $\{L_\lambda\}$ converges in the BW-topology to L if and only if $\text{tr}(L_\lambda(x)T) \to \text{tr}(L(x)T)$ for all $T \in (TC)$ and $x \in X$. But again, since the net is bounded, it is enough to consider $T = R_{h,k}$. □

Let A be a C^*-algebra, S an operator system, M a subspace. We make the following definitions:

$$B_r(M,H) = \{L \in L(M,L(H)): \|L\| \le r\} .$$

$$CB_r(M,H) = \{L \in L(M,L(H)): \|L\|_{cb} \le r\} .$$

$$CP_r(S,H) = \{L \in L(S,L(H)): L \text{ is completely positive}, \|L\| \le r\} .$$

80

$$CP(S,H;P) = \{L \in L(S,L(H)): L \text{ is completely positive, } L(1) = P\} .$$

Theorem 6.4. Let A be a C^*-algebra, let S be a closed operator system, and let M be a closed subspace. Then each of the four above sets is compact in the BW-topology.

Proof. Since the BW-topology is a weak*-topology, the set $B_r(M,H)$ is compact by the Banach-Alaoglu theorem. Since the remaining sets are subsets of this set, it is enough to show that they are closed.

We argue for $CB_r(M,H)$ the rest are similar. Let $\{L_\lambda\}$ be a net in $CB_r(M,H)$ and suppose $\{L_\lambda\}$ converges to L . If $(a_{i,j})$ is in $M_n(M)$, and $x = x_1 \oplus \ldots \oplus x_n$, $y = y_1 \oplus \ldots \oplus y_n$ are in $H \oplus \ldots \oplus H$, then $<L_n((a_{i,j}))x, y> = \lim_\lambda <L_{\lambda n}((a_{i,j}))x, y>$. Hence, $\|L_n((a_{i,j}))\| \leq r \cdot \|(a_{i,j})\|$ for all n and so $\|L\|_{cb} \leq r$.

We're now in a position to prove the main result of the chapter.

Theorem 6.5 (Arveson). Let A be a C^*-algebra, S an operator system, and $\phi: S \to L(H)$ a completely positive map. Then there exists a completely positive map, $\psi: A \to L(H)$, extending ϕ .

Proof. Let F be a finite dimensional subspace of H and let $\phi_F: S \to L(F)$ be the compression of ϕ to F , i.e., $\phi_F(a) = P_F\phi(a)|_F$, where P_F is the projection onto F . Since $L(F)$ is isomorphic to M_n for some n , by Theorem 5.2, there exists a completely positive map $\psi_F: A \to L(F)$ extending ϕ_F . Let $\psi'_F: A \to L(H)$ be defined by setting

$\psi_F'(a)$ equal to $\psi_F(a)$ on F and extending it to be 0 on $F^\perp$.

The set of finite dimensional subspaces of H is a directed set under inclusion, and so $\{\psi_F'\}$ is a net in $CP_r(A,H)$ where $r = \|\phi\|$. Since this latter set is compact, we may choose a subnet which converges to some element $\psi \in CP_r(A,H)$.

We claim that ψ is the desired extension. Indeed, if $a \in S$ and x , y are in H . Let F be the space spanned by x and y . Then for any $F_1 \supseteq F$, $<\phi(a)x, y> = <\psi_{F_1}'(a)x, y>$, and since the set of such F_1 is cofinal, we have that $<\phi(a)x, y> = <\psi(a)x, y>$.

This completes the proof of the theorem. $\qquad\qquad\qquad\square$

Corollary 6.6 (Arveson). Let A be a C^*-algebra, M a subspace with $1 \in M$, and $\phi: M \to L(H)$ a unital, complete contraction. Then there exists a completely positive map $\psi: A \to L(H)$ extending ϕ .

We've seen earlier that positive maps need not have positive extensions (Example 2.13) and that unital, contractive maps need not have contractive extensions (Example 5.10vi) even when the range is finite dimensional. These facts make the above results all the more striking.

A C^*-algebra B is called injective if for every C^*-algebra A and operator system S contained in A , every completely positive map $\phi: S \to B$ can be extended to a completely positive map on all of A . Thus, Theorem 6.5 is the assertion that $L(H)$ is injective. Exercise 6.5 gives an elementary characterization of injective C^*-algebras.

Corollary 6.6 is the basis for a general dilation theory. Let B be a unital C^*-algebra and let A be a subalgebra (not necessarily *-closed) with $1 \in A$. We shall call A an operator algebra. A unital homomorphism

$\rho: A \to L(H)$ is said to have a B-dilation if there exists a Hilbert space K containing H and a unital *-homomorphism $\pi: B \to L(K)$ such that

$$\rho(a) = P_H\pi(a)|_H \qquad \text{for all} \quad a \quad \text{in} \quad A .$$

This definition is motivated in part by the theory of normal ∂X-dilations. Recall that if $T \in L(H)$, then a compact set X is a spectral set for X provided that the homomorphism $\rho: R(X) \to L(H)$ given by $\rho(r) = r(T)$ is contractive. It is clear that T has a normal ∂X-dilation if and only if ρ has a $C(\partial X)$-dilation.

Corollary 6.7 (Arveson). Let A be an operator algebra contained in the C^*-algebra B, let $\rho: A \to L(H)$ be a unital homomorphism, and let $\tilde{\rho}: A + A^* \to L(H)$ be the positive extension of ρ. Then the following are equivalent:

    i) ρ has a B-dilation.

    ii) ρ is completely contractive.

    iii) $\tilde{\rho}$ is completely positive.

Moreover, in this case there exists a B-dilation $\pi: B \to L(K)$ such that $\pi(B)H$ is dense in K.

Proof. We have seen the equivalence of ii) and iii) in Chapter 3. If ρ has a B-dilation, then the map $\phi: B \to L(H)$ defined by

$$\phi(b) = P_H\pi(b)|_H$$

is completely positive and extends ρ, so $\tilde{\rho}$ is completely positive.

Conversely, if $\tilde{\rho}$ is completely positive, then we may extend $\tilde{\rho}$ to a completely positive map $\phi: B \to L(H)$. The Stinespring representation of ϕ gives rise to the B-dilation of ρ .

A minimal Stinespring representation of ϕ will have the property that $\pi(B)H$ is dense in K .   □

A B-dilation with the property that $\pi(B)H$ is dense in K will be called a __minimal__ B-dilation of ρ . These need not be unique (Exercise 6.3).

Let $T \in L(H)$ and let X be a spectral set for T . If the homomorphism $\rho: R(X) \to L(H)$ is completely contractive, then we shall call X a __complete spectral set__ for T .

__Corollary 6.8.__ Let $T \in L(H)$ and let X be a spectral set for T . Then the following are equivalent:

  i) T has a normal ∂X-dilation.

 ii) X is a complete spectral set.

iii) $\tilde{\rho}$ is completely positive.

 iv) $\tilde{\rho}$ has a positive extension to $C(\partial X)$.

Moreover, in this case there is a normal ∂X-dilation N for T such that the smallest closed, reducing subspace for N containing H is K .

__Proof.__ The equivalence of i) - iii) is just Corollary 6.7. If ρ is completely positive, then it has a (completely) positive extension to $C(\partial X)$ by Arveson's extension theorem. Conversely, if $\tilde{\rho}$ does have a positive extension to $C(\partial X)$, then that extension is automatically completely positive by Theorem 3.10, and hence its restriction to S , $\tilde{\rho}$, is completely positive.

If K is the smallest closed, reducing subspace for N containing H ,
then the representation $\pi\colon C(\partial X) \to L(K)$ given by $\pi(z) = N$ has no
closed, reducing subspaces containing H . But this is equivalent to the
requirement that $\pi(C(\partial X))H$ is dense in K , since this latter space
is clearly reducing.   □

A normal ∂X-dilation of T with no closed, reducing subspace containing
H is called a <u>minimal</u>, normal ∂X-dilation of T . Unlike the case of
Sz.-Nagy's minimal, unitary dilation of a contraction (Theorem 4.3), the
minimal, normal ∂X-dilation of an operator need not be unique up to unitary
equivalence. Exercise 6.3 illustrates the difficulty.

A unital homomorphism of an operator algebra A into $L(H)$ which
is one-to-one and contractive is called a <u>representation</u> of A . If A is a
C^*-algebra, then every representation of A is automatically a *-representa-
tion. Corollary 6.7 gives a useful characterization of the completely
contractive representations for it and shows that these representations
can be studied via the representation theory of B . Consequently, it is
important to know when representations are completely contractive.

At present we have no natural source of representations which are not
completely contractive. In general, such examples are hard to construct,
since in some sense they are the pathological cases which are difficult
to fit to a general theory. They can occur, however, even when the algebra
is finite dimensional and the representation is on a finite dimensional
space. An example of a subalgebra of M_3 and a representation of it on
$\mathbb{C}^3$ which is not completely contractive is given in [3].

An example of considerable importance to the theory of dilations of
contractions was given by Parrott [83].

6.9 Parrott's Example. Let U and V be contractions in $L(H)$ such that U is unitary and U and V don't commute. We define commuting contractions on $L(H \oplus H)$ by setting

$$T_1 = \begin{bmatrix} 0 & 0 \\ I & 0 \end{bmatrix}, \quad T_2 = \begin{bmatrix} 0 & 0 \\ U & 0 \end{bmatrix}, \quad T_3 = \begin{bmatrix} 0 & 0 \\ V & 0 \end{bmatrix}.$$

Let P_3 be the algebra of polynomials in 3 variables, z_1, z_2, z_3, regarded as a subalgebra of $C(\Pi^3)$, where Π^3 is the 3 torus. We claim that the homomorphism

$$\rho: P_3 \to L(H \oplus H) \text{ defined by } \rho(z_i) = T_i, \quad i = 1, 2, 3,$$

is contractive but not completely contractive.

To see that ρ is contractive, let $p(z_1, z_2, z_3)$ be an arbitrary element of P_3 and write

$$p(z_1, z_2, z_3) = a_o + a_1 z_1 + a_2 z_2 + a_3 z_3 + q(z_1, z_2, z_3),$$

where $q(z_1, z_2, z_3)$ contains all the higher order terms of p. We have that

$$\rho(p) = \begin{bmatrix} a_o & 0 \\ a_1 I + a_2 U + a_3 V & a_o \end{bmatrix},$$

since $T_i \cdot T_j = 0$ for all i and j.

Let $x = x_1 \oplus x_2$ and $y = y_1 \oplus y_2$ be arbitrary unit vectors in $H \oplus H$, and calculate

$$|<\rho(p)x, y>| = |a_o<x_1 \cdot y_1> + <(a_1 I + a_2 U + a_3 V)x_1, y_2> + a_o<x_2, y_2>|$$

$$\leq |a_o|\|x_1\| \ \|y_1\| + (|a_1| + |a_2| + |a_3|)\|x_1\| \ \|y_2\| + |a_o|\|x_2\| \ \|y_2\|$$

$$= \left\langle \begin{bmatrix} |a_o| & 0 \\ |a_1| + |a_2| + |a_3| & |a_o| \end{bmatrix} \begin{bmatrix} \|x_1\| \\ \|x_2\| \end{bmatrix}, \begin{bmatrix} \|y_1\| \\ \|y_2\| \end{bmatrix} \right\rangle .$$

Thus, we have that

$$\|\rho(p)\| \leq \left\| \begin{bmatrix} |a_o| & 0 \\ |a_1| + |a_2| + |a_3| & |a_o| \end{bmatrix} \right\|$$

where the latter matrix is an element of M_2 .

But by Exercise 2.11,

$$\left\| \begin{bmatrix} |a_o| & 0 \\ |a_1| + |a_2| + |a_3| & |a_o| \end{bmatrix} \right\| \leq \inf_r \{ \| |a_o| + (|a_1|+|a_2|+|a_3|)z + r(z) \| \},$$

where $r(z)$ is an arbitrary polynomial whose lowest order term is at least degree 2 , and the latter norm is the supremum norm over the unit circle. Let λ_o , λ_1 , λ_2 , λ_3 be numbers of modulus 1 such that $\lambda_o a_o$, $\lambda_1 a_1$, $\lambda_2 a_2$, $\lambda_3 a_3$ are positive, then

$$\|\rho(p)\| \leq \left\| \begin{bmatrix} |a_o| & 0 \\ |a_1| + |a_2| + |a_3| & |a_o| \end{bmatrix} \right\|$$

$$\leq \| |a_o| + (|a_1|+|a_2|+|a_3|)z + \lambda_o q(\bar{\lambda}_o \lambda_1 z, \bar{\lambda}_o \lambda_2 z, \bar{\lambda}_o \lambda_3 z) \|$$

$$= \|p(\bar{\lambda}_o \lambda_1 z, \bar{\lambda}_o \lambda_2 z, \bar{\lambda}_o \lambda_3 z) \|$$

$$\leq \|p(z_1, z_2, z_3) \|$$

where the 3rd and 4th norms are the supremum over Π and the last norm is the supremum over Π^3 . Thus, ρ is contractive.

Now assume that ρ is completely contractive. Consider an element of $M_n(P_3)$ of the form $(a_{i,j}z_1 + b_{i,j}z_2 + c_{i,j}z_3)$. Its image under ρ_n is

$$
\begin{bmatrix}
0 & 0 & & & 0 & 0 \\
a_{11}I + b_{11}U + c_{11}V & 0 & \cdots & & a_{1n}I + b_{1n}U + c_{1n}V & 0 \\
\cdot & \cdot & & & \cdot & \\
\cdot & \cdot & & \cdot & \cdot & \\
\cdot & \cdot & & & \cdot & \\
0 & 0 & \cdots & & 0 & 0 \\
a_{n1}I + b_{n1}U + c_{n1}V & 0 & & & a_{nn}I + b_{nn}U + c_{nn}V & 0
\end{bmatrix},
$$

which is an operator acting on the direct sum of n copies of $(H \oplus H)$. Think of this space as $(H_{11} \oplus H_{21}) \oplus \cdots \oplus (H_{1n} \oplus H_{2n})$ where each $H_{i,j} = H$. If we re-order the spaces by writing $(H_{11} \oplus \cdots \oplus H_{1n}) \oplus (H_{21} \oplus \cdots \oplus H_{2n})$, then the matrix of this same operator will be

$$
\begin{bmatrix}
0 & \cdots & & \cdots & 0 & 0 & \cdots & 0 \\
\cdot & & & & & \cdot & & \\
\cdot & & & & \cdot & \cdot & & \\
\cdot & & & & & \cdot & & \\
0 & \cdots & & \cdots & 0 & 0 & \cdots & 0 \\
a_{11}I + b_{11}U + c_{11}V & \cdots & & a_{1n}I + b_{1n}U + c_{1n}V & 0 & \cdots & 0 \\
& & & \cdot & & \cdot & & \\
& & \cdot & & & \cdot & & \\
& & & \cdot & & \cdot & & \\
a_{n1}I + b_{n1}U + c_{n1}V & \cdots & & a_{nn}I + b_{nn}U + c_{nn}V & 0 & \cdots & 0
\end{bmatrix}
$$

Hence, the operator matrix in the (2,1)-block of the above matrix has

norm bounded by

$$\|\rho_n((a_{i,j}z_1 + b_{i,j}z_2 + c_{i,j}z_3))\| \le \sup\ \{\|(a_{i,j}z_1 + b_{i,j}z_2 + c_{i,j}z_3)\|:\ |z_1| = |z_2| = |z_3| = 1\}\ .$$

But this latter supremum is equal to

$$\sup\ \{\|(a_{i,j} + b_{i,j}w_1 + c_{i,j}w_2)\|:\ |w_1| = |w_2| = 1\}\ .$$

Let M be the subspace of $C(\pi^2)$ spanned by 1, w_1, and w_2, where

w_1 and w_2 are the coordinate functions.

The above discussion shows that if ρ is completely contractive,

then the map $\psi: M \to L(H)$ defined by $\psi(1) = I$, $\psi(w_1) = U$, and

$\psi(w_2) = V$ is completely contractive. By Arveson's extension theorem,

we may extend ψ to a completely positive map on $C(\pi^2)$ and then obtain

a Stinespring representation (π, S, K) with S an isometry and

$\pi: C(\pi^2) \to L(K)$ a *-homomorphism.

Identifying SH and H we have that

$$\psi(\cdot) = P_H\pi(\cdot)|_H\ .$$

But since $\pi(w_1)$ and $U = \psi(w_1)$ are both unitary, this implies that H

reduces $\pi(w_1)$. Finally, since $\pi(w_1)$ and $\pi(w_2)$ commute, and H

reduces $\pi(w_1)$, this implies that the compression of $\pi(w_1)$ to H commutes

with the compression of $\pi(w_2)$ to H. That is, that U commutes with

V, a contradiction. Therefore, ρ is not completely contractive.

If one chooses U and V to both be unitaries, then it is possible

to show that ρ is not even 4-contractive (Exercise 6.9).

Another remark to be made is that there exist U and V in M_2

satisfying the above conditions, and hence there are three commuting contractions in M_4 for which ρ is a contraction but not a complete contraction.

Now that we've seen an example of a commutative operator algebra for which not every representation is completely contractive, we would like to present an example in the opposite extreme. That is, some highly noncommutative operator algebras for which every representation is completely contractive.

Proposition 6.10 (McAsey-Muhly). Let A be the algebra of upper triangular matrices in M_n, i.e.,

$$A = \{(a_{i,j}): a_{i,j} = 0 \text{ for } i > j\} .$$

Then every representation of A is completely contractive.

Proof. Let $\rho: A \to L(H)$ be a representation. Since $A + A^* = M_n$, the map $\phi: M_n \to L(H)$ defined by $\phi(A + B^*) = \rho(A) + \rho(B)^*$ is positive. We must prove that ϕ is completely positive.

Let $\{E_{i,j}\}$ be the standard matrix units for M_n. Since ρ is unital and contractive, $\tilde{\rho}$ is well-defined and positive. Thus, $\{\rho(E_{i,i})\}$ will be orthogonal projections which sum to the identity. Let H_i be the space that $\rho(E_{i,i})$ projects onto so $H = H_1 \oplus \dots \oplus H_n$.

Since $\rho(E_{i,j}) = \rho(E_{i,i}) \rho(E_{i,j}) \rho(E_{j,j})$, there will exist operators $T_{i,j}: H_j \to H_i$ such that the operator matrix of $\rho(E_{i,j})$ relative to the above decomposition of H is $T_{i,j}$ in the (i,j)-th position and 0 elsewhere. Set $T_i = T_{i,i+1}$ and note that for $i < j$,

$$T_{i,j} = T_i \cdot T_{i+1} \cdots T_j .$$

By Theorem 3.12, to prove that ϕ is completely positive, it is sufficient to prove that $(\phi(E_{i,j}))$ is a positive operator on the direct sum of n-copies of H, $H' \oplus \ldots \oplus H^n$. Decomposing each H^i as $H^i_1 \oplus \ldots \oplus H^i_n$, we have that $(\phi(E_{i,j}))$ is represented as an $n^2 \times n^2$ operator matrix with each $n \times n$ block having only one non-zero entry.

Reorder the subspaces such that H^1_1, H^2_2, $\ldots$, H^n_n are listed first and the remaining spaces occur in any order.

The operator matrix for $(\phi(E_{i,j}))$ with respect to this reordering is $(T_{i,j})$ in the first $n \times n$ entries and 0 elsewhere.

Thus, it will be sufficient to prove that $(T_{i,j})$ is positive. Set

$$
R = \begin{bmatrix}
0 & T_1 & 0 & . & . & . & 0 \\
 & & & . & & . & . \\
0 & 0 & . & T_2 . & & . & 0 \\
. & & & . & . & . & \\
. & & & & . & . & T_{n-1} \\
. & & & & & . & \\
0 & . & . & . & . & . & 0
\end{bmatrix} \quad ,
$$

then $R \leq 1$, $R^{n+1} = 0$, and $(T_{i,j}) = (I - R)^{-1} + (I - R^*)^{-1} - I$. Consequently, as in the proof of Theorem 2.6, $(T_{i,j})$ is positive. $\square$

Since so many positive maps on M_n are not completely positive, it is somewhat surprising that every map induced by a representation of A is completely positive.

NOTES

The BW-topology was introduced in [3] where the proofs of the extension theorems also appear, as well as some material on B-dilations.

Parrott [83] originally proved that the set of operators in Example 6.9 did not have a set of commuting unitary dilations, i.e., that $\rho(z_1, z_2, z_3) \rightarrow \rho(T_1, T_2, T_3)$ does not have a $C(\Pi^3)$-dilation by examining spatial relations among the unitaries. By Corollary 6.7, this is equivalent to the fact that Parrott's example yields a contractive homomorphism that is not completely contractive. We've avoided Parrott's spatial proof in an attempt to cast a different perspective. It is interesting to note that when V is also unitary, a direct proof of the non-complete contractivity of the homomorphism simplifies (Exercise 6.9), as does the spatial proof [114]. We do not know if the homomorphism is not 2-contractive for arbitrary V.

Let us re-interpret the material on commuting contractions in light of Corollary 6.7. Ando's theorem [2] shows that every pair of commuting contractions induces a completely contractive homomorphism of the bidisk algebra, $P(\mathbb{D}^2)$. So, in particular, every contractive homomorphism of that algebra is completely contractive. Parrott's example shows that $P(\mathbb{D}^3)$ differs in this respect; contractive homomorphisms need not be completely contractive. The internal properties of the algebra $P(\mathbb{D}^3)$ that lead to this difference do not seem to be understood. Neither do the internal properties of the bidisk algebra. For example, it is not known if every unital contraction (including non-homomorphisms) on $P(\mathbb{D}^2)$ is completely contractive. This is the case for the disk algebra, since it is a Dirichlet algebra.

The results of Crabbe-Davies [31] and Varopolos [119], that not every set of $n \geq 3$ commuting contractions need satisfy von Neumann's inequality, are more of a property of Hilbert space than of the polydisc algebras. Gaspar and Racz [43] have shown that every set of cyclically commuting contractions has a set of cyclically commuting, unitary dilations. If given a C^*-algebraic interpretation, their result should add some insight.

Choi-Effros [24] prove that for a von Neumann algebra A, semi-discreteness (Exercise 6.14) is equivalent to the existence of a net of finite rank, completely positive maps, $R_\lambda : A \to A$, such that $R_\lambda(a)$ converges to a σ-weakly for all a. Connes [29] proves that semi-discreteness and injectivity are equivalent for von Neumann algebras.

Proposition 5.10 is adapted from McAsey-Muhly [73].

EXERCISES

6.1 Let X and Y be Banach spaces and for $f \in X^*$, $g \in Y^*$, define $L_{f,g} \in L(X,Y^*)$ by $L_{f,g}(x) = f(x)q$. By considering these operators, prove that the map $j: X \otimes Y \to L(X,Y^*)^*$ defined in this chapter has the following properties:

   i) j is linear.

   ii) $\|j(x \otimes y)\| = \|x\| \|y\|$ (such norms are called cross-norms).

   iii) j is one-to-one.

Conclude that Z can be identified with the completion of $X \otimes Y$ with respect to the above cross-norm.

The above cross-norm is called the injective cross-norm, and the completion of $X \otimes Y$ with respect to this norm is called the injective tensor product of X and Y.

6.2 Let $R_{h,k}$ be the operator defined by $R_{h,k}(x) = <x,k>h$. Show that $R_{h,k}$ is trace class, that $\mathrm{tr}\,(AR_{h,k}) = <Ah,k>$, and that the linear span of $\{R_{h,k}: h, k \in H\}$ is dense in (TC) in the trace norm.

6.3 Let A be an operator algebra contained in the C^*-algebra B, let $\rho: A \to L(H)$ be a completely contractive, unital homomorphism, and let $\pi_i: B \to L(K_i)$, $i = 1$, 2, define minimal B-dilations of ρ. Define completely positive maps, $\phi_i: B \to L(H)$, by $\phi_i(b) = P_H\pi_i(b)|_H$, $i = 1$, 2.

    i) Show that there exists a unitary $U: K_1 \to K_2$ with $Uh = h$ for h in H, and $U^*\pi_2(b)U = \pi_1(b)$ if and only if $\phi_1 = \phi_2$. Such dilations are called <u>unitarily</u> <u>equivalent</u>.

    ii) Show that there is a one-to-one correspondence between unitarily equivalent, minimal B-dilations of ρ and completely positive extensions of ρ to B.

    iii) Show that the set of completely positive extensions of ρ is a compact, convex set in the BW-topology on $CP(B,H)$.

6.4 (Extension of Bimodule Maps) Let A, C be C^*-algebras, let S be an operator system, and suppose that $C \subseteq S \subseteq A$. If $C \subseteq L(H)$, then $\phi: S \to L(H)$ is a C-bimodule map provided $\phi(c_1ac_2) = c_1\phi(a)c_2$. Prove that if $\phi: S \to L(H)$ is a completely positive C-bimodule map, then ϕ has a completely positive, C-bimodule extension to A.

6.5 Let $B \subseteq L(H)$ be unital. Prove that B is injective if and only if there exists a completely positive map $\phi: L(H) \to B$ such that $\phi(b) = b$ for all b in B. Show that ϕ is necessarily a B-bimodule map. A map with the above properties is called a <u>completely positive conditional expectation</u>.

6.6 (Sarason) Let $A \subseteq L(K)$ be an algebra, and let H be a subspace of K such that $A \to P_HA|_H$ is a homomorphism on A. Prove that there are subspaces H_1 and H_2 of K, both of which are invariant for every element of A, such that $H_1 \oplus H = H_2$. Conversely, show that if H_1 and H_2 are invariant for A and $H_1 \oplus H = H_2$, then compression onto H is a homomorphism on A. Such a subspace is said to be <u>semi-invariant</u> for A.

6.7 Let $Q = (q_{i,j})$ be the element of M_n given by $q_{i,i+1} = 1$, $q_{i,j} = 0$ for all other (i,j), and let $N \in L(H)$ satisfy $\|N\| \le 1$, $N^n = 0$. Prove that $Q \to N$ defines a completely contractive representation of the algebra generated by Q.

6.8 Prove that B is injective if and only if $M_n(B)$ is injective for some n.

6.9 In Parrott's example, show that if U and V are both unitaries that don't commute, then ρ is not 4-contractive.

$$\left[\text{HINT: Consider} \begin{bmatrix} U^{\star} & V \\ V^{\star} & -U \end{bmatrix} \right].$$

6.10 (Parrott) Prove that there exist three commuting operators. T_1, T_2, and T_3, with $\|T_i\| < 1$ such that von Neumann's inequality holds, but which do not have a commuting unitary dilation.

6.11 Prove that for every $n \ge 3$, there exist n commuting operators such that $\|T_i\| < 1$, $i = 1, \ldots, n$, von Neumann's inequality holds, but which do not have commuting unitary dilations.

6.12 Construct a unital, isometric homomorphism of $P(\mathbb{D}^3)$ that is not completely isometric.

6.13 Let $A \subseteq L(H)$ be a unital C^*-algebra. If A is closed in the weak operator topology, then A is called a von Neumann algebra. Prove that every von Neumann algebra has a pre-dual. The induced weak*-topology on A is called the σ-weak topology.

6.14 Let A be a von Neumann algebra. If there exists a net of σ-weakly continuous, completely positive maps, $\phi_\lambda : A \to M_{n_\lambda}$, $\psi_\lambda : M_{n_\lambda} \to A$, with $\phi_\lambda(1) = 1$, $\psi_\lambda(1) = 1$, such that $\psi_\lambda \circ \phi_\lambda(a) \to a$, σ-weakly for all a in A, then A is called semi-discrete. Prove that every semi-discrete von Neumann algebra is injective.

7 Completely bounded maps

In this chapter, we extend many of the results obtained in previous chapters concerning completely positive maps to the completely bounded maps. Our main technique is to realize completely bounded maps as the off-diagonal corners of completely positive maps.

Let us consider for a moment $M_m(M_n(A))$. For a C^*-algebra A , a typical element of this algebra is of the form $A = (A_{i,j})_{i,j=1}^m$, where each $A_{i,j}$ is in $M_n(A)$. Thus, $A_{i,j} = (a_{i,j,k,\ell})_{k,\ell=1}^n$, with $a_{i,j,k,\ell}$ in A . Setting $B_{k,\ell} = (a_{i,j,k,\ell})_{i,j=1}^m$, we obtain an element of $M_m(A)$ and thus, $B = (B_{k,\ell})_{k,\ell=1}^n$ is in $M_n(M_m(A))$. Now, $M_m(M_n(A))$ and $M_n(M_m(A))$ are both isomorphic to $M_{nm}(A)$ by just deleting the extra parentheses. With these identifications, A and B are unitarily equivalent elements of $M_{nm}(A)$, in fact, the unitary is just a permutation matrix.

To see this, note that when we regard A as an element of $M_{mn}(A)$, say $A = (c_{s,t})_{s,t=1}^{mn}$, then $c_{s,t} = a_{i,j,k,\ell}$, where $s = n(i - 1) + k$, $t = n(j - 1) + \ell$. While, if we regard B as an element of $M_{mn}(A)$, say $B = (d_{s,t})_{s,t=1}^{mn}$, then $d_{s,t} = a_{i,j,k,\ell}$, where $s = m(k - 1) + i$, $t = m(\ell - 1) + j$.

Since the above operation for passing from $M_m(M_n(A))$ to $M_n(M_m(A))$ is just a permutation, it is a *-isomorphism. We shall refer to this *-isomorphism as the <u>canonical shuffle</u>. It is important to note that since the canonical shuffle is a *-isomorphism, it preserves norm and positivity.

Note that we've already encountered this canonical shuffle in 6.9

and 6.10.

It is also useful to understand this canonical shuffle in the tensor notation. Let $\{E_{i,j}\}_{i,j=1}^m$ and $\{F_{k,\ell}\}_{k,\ell=1}^n$ denote the standard matrix units for M_m and M_n, respectively. Our element A of $M_m(M_n(A)) \cong (A \otimes M_n) \otimes M_m$ is just $A = \Sigma_{i,j=1}^m A_{i,j} \otimes E_{i,j}$, where each $A_{i,j}$ is in $A \otimes M_n$ and has the form $A_{i,j} = \Sigma_{k,\ell=1}^n a_{i,j,k,\ell} \otimes F_{k,\ell}$. Thus, $A = \Sigma_{i,j=1}^m \Sigma_{k,\ell=1}^n a_{i,j,k,\ell} \otimes F_{k,\ell} \otimes E_{i,j}$. On the other hand, $B_{k,\ell} = \Sigma_{i,j=1}^m a_{i,j,k,\ell} \otimes E_{i,j}$, so that $B = \Sigma_{k,\ell=1}^n B_{k,\ell} \otimes F_{k,\ell} = \Sigma_{k,\ell=1}^n \Sigma_{i,j=1}^m a_{i,j,k,\ell} \otimes E_{i,j} \otimes F_{k,\ell}$, which lies in $(A \otimes M_m) \otimes M_n \cong M_n(M_m(A))$. When we regard A and B as elements of $M_{mn}(A)$, then we see that B is the image of A under the following string of isomorphisms; $M_{mn}(A) \cong M_m(M_n(A)) \cong (A \otimes M_n) \otimes M_m \cong A \otimes (M_n \otimes M_m) \cong A \otimes (M_m \otimes M_n) \cong (A \otimes M_m) \otimes M_n \cong M_n(M_m(A)) \cong M_{mn}(A)$. Most of these isomorphisms are so canonical that it is customary to ignore them. Setting $A = \mathbb{C}$, it is interesting to note that when we identify $M_{mn} = M_m(M_n) = M_n \otimes M_m$ and $M_{mn} = M_n(M_m) = M_m \otimes M_n$, then it is only the isomorphism between $M_m \otimes M_n$ and $M_n \otimes M_m$ which leads to the permutation in M_{mn}.

The following lemma is central to this section and introduces the off-diagonal technique.

Lemma 7.1. Let A, B be C^*-algebras with unit 1, let M be an operator space in A, and let $\phi: M \to B$. Define an operator system $S \subseteq M_2(A)$ by

$$
S = \left\{ \begin{bmatrix} \lambda 1 & a \\ b^\star & \mu 1 \end{bmatrix} : \lambda, \mu \in \mathbb{C}, \ a,b \in M \right\},
$$

and $\Phi: S \rightarrow M_2(B)$ via

$$\Phi\left(\left[\begin{array}{cc} \lambda 1 & a \\ b^* & \mu 1 \end{array}\right]\right) = \left[\begin{array}{cc} \lambda 1 & \phi(a) \\ \phi(b)^* & \mu 1 \end{array}\right] .$$

If ϕ is completely contractive, then Φ is completely positive.

Proof. Let $(S_{i,j})$ be in $M_n(S)$, say

$$S_{i,j} = \left[\begin{array}{cc} \lambda_{i,j} & a_{i,j} \\ b^*_{i,j} & \mu_{i,j} \end{array}\right] .$$

Since $M_n(S)$ is a subspace of $M_n(M_2(A))$, if we perform the canonical shuffle, then $(S_{i,j})$ becomes an element of $M_2(M_n(A))$. Indeed, if we set $H = (\lambda_{i,j})$, $A = (a_{i,j})$, $B = (b_{j,i})$, $K = (\mu_{i,j})$, then the image of $(S_{i,j})$ under the canonical shuffle is

$$(1) \qquad \left[\begin{array}{cc} H & A \\ B^* & K \end{array}\right] .$$

Similarly, after the canonical shuffle, $\Phi_n((S_{i,j}))$ becomes

$$(2) \qquad \left[\begin{array}{cc} H & \phi_n(A) \\ \phi_n(B)^* & K \end{array}\right] .$$

Thus, to prove that Φ is completely positive, it is sufficient to prove that for all n , if (1) is positive, then (2) is positive.

Now, if (1) is positive, then $A = B$ and H and K must be

positive. Fix $\varepsilon > 0$ and set $H_\varepsilon = H + \varepsilon I$, $K_\varepsilon = K + \varepsilon I$, so that H_ε
and K_ε are positive and invertible. We have that

$$
\begin{bmatrix} I & H_\varepsilon^{-\frac{1}{2}}AK_\varepsilon^{-\frac{1}{2}} \\ K_\varepsilon^{-\frac{1}{2}}A^*H_\varepsilon^{-\frac{1}{2}} & I \end{bmatrix} = \begin{bmatrix} H_\varepsilon^{-\frac{1}{2}} & 0 \\ 0 & K_\varepsilon^{-\frac{1}{2}} \end{bmatrix} \begin{bmatrix} H_\varepsilon & A \\ A^* & K_\varepsilon \end{bmatrix} \begin{bmatrix} H_\varepsilon^{-\frac{1}{2}} & 0 \\ 0 & K_\varepsilon^{-\frac{1}{2}} \end{bmatrix}
$$

is positive, and consequently by Lemma 3.1,

$$
\| H_\varepsilon^{-\frac{1}{2}}AK_\varepsilon^{-\frac{1}{2}} \| \le 1 .
$$

Also, $\phi_n(H_\varepsilon^{-\frac{1}{2}}AK_\varepsilon^{-\frac{1}{2}}) = H_\varepsilon^{-\frac{1}{2}}\phi_n(A)K_\varepsilon^{-\frac{1}{2}}$ (Exercise 7.1), and so

$$
\begin{bmatrix} H_\varepsilon & \phi_n(A) \\ \phi_n(A)^* & K_\varepsilon \end{bmatrix} = \begin{bmatrix} H_\varepsilon^{\frac{1}{2}} & 0 \\ 0 & K_\varepsilon^{\frac{1}{2}} \end{bmatrix} \begin{bmatrix} I & \phi_n(H^{-\frac{1}{2}}AK^{-\frac{1}{2}}) \\ \phi_n(H^{-\frac{1}{2}}AK^{-\frac{1}{2}})^* & I \end{bmatrix} \begin{bmatrix} H_\varepsilon^{\frac{1}{2}} & 0 \\ 0 & K_\varepsilon^{\frac{1}{2}} \end{bmatrix} .
$$

But, since ϕ is completely contractive, $\| \phi_n(H^{-\frac{1}{2}}AK^{-\frac{1}{2}}) \| \le 1$ and so by
another application of Lemma 3.1, the middle term on the right of the
above equation is positive. Thus, the left hand side is positive for all
ε, and so (1) is positive.

Consequently, Φ is completely positive, which completes the proof. □

We can now prove an extension theorem for completely bounded maps.

Theorem 7.2. Let A be a unital C^*-algebra, M a subspace of A,
and let $\phi: M \to L(H)$ be completely bounded. Then there exists a
completely bounded map, $\psi: A \to L(H)$, which extends ϕ, with $\| \phi \|_{cb} = \| \psi \|_{cb}$.

Proof. We may assume without loss of generality that $\| \phi \|_{cb} = 1$.

Let S and Φ be as in Lemma 7.1. Since Φ is completely positive by Arveson's extension theorem, there exists a completely positive $\Psi: M_2(A) \to M_2(L(H)) = L(H \oplus H)$ extending Φ.

Define ψ via

$$\psi\left(\begin{bmatrix} 0 & a \\ 0 & 0 \end{bmatrix}\right) = \begin{bmatrix} * & \psi(a) \\ * & * \end{bmatrix}.$$

Clearly, ψ is linear and since Ψ extends Φ, ψ extends ϕ. Also, since Ψ is unital,

$$\|\psi(a)\| \leq \left\| \Psi \begin{bmatrix} 0 & a \\ 0 & 0 \end{bmatrix} \right\| \leq \|\Psi\| \cdot \|a\| \leq \|a\|,$$

and so ψ is contractive.

To see that ψ is completely contractive, let $A = (a_{i,j})$ be in $M_n(A)$, then

$$\Psi_n\left(\left[\begin{bmatrix} 0 & a_{i,j} \\ 0 & 0 \end{bmatrix}\right]^n_{i,j=1}\right) = \left[\begin{bmatrix} * & \psi(a_{i,j}) \\ * & * \end{bmatrix}\right]^n_{i,j=1}$$

After performing the canonical shuffle, the right-hand side becomes

$$\begin{bmatrix} * & \psi_n(A) \\ * & * \end{bmatrix},$$

where the asterisk now indicates an $n \times n$ rather than 1×1 entry.

Thus, by the same inequality as above,

$$\|\psi_n(A)\| \leq \left\| \left[\begin{bmatrix} 0 & a_{i,j} \\ 0 & 0 \end{bmatrix} \right]_{i,j=1}^{n} \right\| .$$

But after a canonical shuffle, $\left[\begin{bmatrix} 0 & a_{i,j} \\ 0 & 0 \end{bmatrix} \right]_{i,j=1}^{n}$ becomes $\begin{bmatrix} 0 & A \\ 0 & 0 \end{bmatrix}$.

Thus, $\|\psi_n(A)\| \leq \|A\|$, which is what we needed to prove. □

If M is an operator space in the C^*-algebra A , B is another C^*-algebra, and $\phi: M \to B$ is a linear map, we set $M^* = \{a: a^* \in M\}$ and define a linear map

$$\phi^*: M^* \to B , \text{ via } \phi^*(a) = \phi(a^*)^* .$$

When $M = M^*$, we set

$$\text{Re } \phi = (\phi + \phi^*)/2 , \quad \text{Im } \phi = (\phi - \phi^*)/2i ,$$

so that $\text{Re } \phi$, $\text{Im } \phi$ are self-adjoint, linear maps with $\phi = \text{Re } \phi + i \text{ Im } \phi$. We note that some care is needed, since, in general, $(\text{Re } \phi)(a) \neq \text{Re}(\phi(a))$, but they are equal when $a = a^*$.

The above Lemma also yields the following decomposition theorem for completely bounded maps.

Theorem 7.3. Let A be a C^*-algebra with unit and let $\phi: A \to L(H)$ be completely bounded. Then there exists completely positive maps, $\phi_i: A \to L(H)$, with $\|\phi_i\|_{cb} = \|\phi\|_{cb}$, $i = 1, 2$, such that the map $\Phi: M_2(A) \to L(H \oplus H)$ given by

$$\Phi\left(\begin{bmatrix} a & b \\ c & d \end{bmatrix}\right) = \begin{bmatrix} \phi_1(a) & \phi(b) \\ \phi^*(c) & \phi_2(d) \end{bmatrix}$$

is completely positive. Moreover, if $\|\phi\|_{cb} \leq 1$, then we may take $\phi_1(1) = \phi_2(1) = 1$.

Proof. Clearly, we may assume that $\|\phi\|_{cb} = 1$. Applying Lemma 7.1 with $M = A$, we obtain a completely positive map $\Phi: S \to L(H \oplus H)$ where $S \subseteq M_2(A)$.

By Arveson's extension theorem, Φ extends to a completely positive map on all of $M_2(A)$, which we still denote by Φ. Since $\begin{bmatrix} 0 & b \\ c & 0 \end{bmatrix}$ is in S, by the definition of Φ,

$$\Phi\left(\begin{bmatrix} 0 & b \\ c & 0 \end{bmatrix}\right) = \begin{bmatrix} 0 & \phi(b) \\ \phi(c^*)^* & 0 \end{bmatrix} = \begin{bmatrix} 0 & \phi(b) \\ \phi^*(c) & 0 \end{bmatrix} .$$

Now, let p be positive, $p \leq 1$, then since

$$\begin{bmatrix} p & 0 \\ 0 & 0 \end{bmatrix} \leq \begin{bmatrix} 1 & 0 \\ 0 & 0 \end{bmatrix} ,$$

we have that

$$\begin{bmatrix} 0 & 0 \\ 0 & 0 \end{bmatrix} \leq \Phi\left(\begin{bmatrix} p & 0 \\ 0 & 0 \end{bmatrix}\right) \leq \Phi\left(\begin{bmatrix} 1 & 0 \\ 0 & 0 \end{bmatrix}\right) = \begin{bmatrix} 1 & 0 \\ 0 & 0 \end{bmatrix} .$$

A straightforward calculation shows that these inequalities taken together imply that

$$\Phi\left(\left[\begin{array}{cc} p & 0 \\ 0 & 0 \end{array}\right]\right) = \left[\begin{array}{cc} * & 0 \\ 0 & 0 \end{array}\right] .$$

Since A is the span of its positive elements, there must exist a linear map $\phi_1 : A \rightarrow L(H)$ such that

$$\Phi\left(\left[\begin{array}{cc} a & 0 \\ 0 & 0 \end{array}\right]\right) = \left[\begin{array}{cc} \phi_1(a) & 0 \\ 0 & 0 \end{array}\right] .$$

By an argument similar to the proof in Theorem 7.2, that ψ is completely contractive, one obtains that ϕ_1 is completely positive.

By analogous arguments, one obtains a completely positive map $\phi_2 : A \rightarrow L(H)$ satisfying

$$\Phi\left(\left[\begin{array}{cc} 0 & 0 \\ 0 & d \end{array}\right]\right) = \left[\begin{array}{cc} 0 & 0 \\ 0 & \phi_2(d) \end{array}\right] .$$

Thus, we see that any completely positive map Φ obtained via extension from S to $M_2(A)$ has the desired form,

$$\Phi\left(\left[\begin{array}{cc} a & b \\ c & d \end{array}\right]\right) = \Phi\left(\left[\begin{array}{cc} 0 & b \\ c & 0 \end{array}\right]\right) + \Phi\left(\left[\begin{array}{cc} a & 0 \\ 0 & d \end{array}\right]\right) = \left[\begin{array}{cc} \phi_1(a) & \phi(b) \\ \phi^*(c) & \phi_2(d) \end{array}\right] ,$$

for some ϕ_1 and ϕ_2.

Since $\phi_1(1) = \phi_2(1) = 1$, we have that $\|\phi_1\|_{cb} = \|\phi_2\|_{cb} = \|\phi\|_{cb}$. This completes the proof of the theorem. ☐

The above decomposition leads readily to a generalization of Stinespring's representation.

104

Theorem 7.4. Let A be a C^*-algebra with unit and let $\phi: A \to L(H)$ be a completely bounded map. Then there exists a Hilbert space K, a *-homomorphism $\pi: A \to L(K)$, and bounded operators $V_i: H \to K$, $i = 1, 2$, with $\|\phi\|_{cb} = \|V_1\| \cdot \|V_2\|$ such that

$$\phi(a) = V_1^* \pi(a) V_2$$

for all $a \in A$. Moreover, if $\|\phi\|_{cb} = 1$, then V_1 and V_2 may be taken to be isometries.

Proof. Clearly, we may assume that $\|\phi\|_{cb} = 1$. Let ϕ_1, ϕ_2, and Φ be as in Theorem 7.3. Let (π_1, V, K_1) be a minimal, Stinespring representation for Φ, and note that since Φ is unital, V may be taken to be an isometry and π_1 to be unital.

Since $M_2(A)$ contains a copy of M_2, the Hilbert space K_1 may be decomposed as $K_1 = K \oplus K$ in such a way that the *-homomorphism $\pi_1: M_2(A) \to L(K \oplus K)$ has the form

$$\pi_1 \left(\begin{bmatrix} a & b \\ c & d \end{bmatrix} \right) = \begin{bmatrix} \pi(a) & \pi(b) \\ \pi(c) & \pi(d) \end{bmatrix} ,$$

where $\pi: A \to L(K)$ is a unital *-homomorphism (Exercise 7.3).

Thus, we have that $V: H \oplus H \to K \oplus K$ is an isometry, and

$$\begin{bmatrix} \phi_1(a) & \phi(b) \\ \phi^*(c) & \phi_2(d) \end{bmatrix} = V^* \begin{bmatrix} \pi(a) & \pi(b) \\ \pi(c) & \pi(d) \end{bmatrix} V .$$

For h in H,

$$\begin{bmatrix} h \\ 0 \end{bmatrix} = \begin{bmatrix} \phi_1(1) & 0 \\ 0 & 0 \end{bmatrix} \begin{bmatrix} h \\ 0 \end{bmatrix} = V^* \begin{bmatrix} \pi(1) & 0 \\ 0 & 0 \end{bmatrix} V \begin{bmatrix} h \\ 0 \end{bmatrix} = V^* \begin{bmatrix} 1_K & 0 \\ 0 & 0 \end{bmatrix} V \begin{bmatrix} h \\ 0 \end{bmatrix} ,$$

and since V is an isometry, we must have that

$$V \begin{bmatrix} h \\ 0 \end{bmatrix} = \begin{bmatrix} \star \\ 0 \end{bmatrix} .$$

Thus, there is a linear map $V_1: H \to K$ such that $V \begin{bmatrix} h \\ 0 \end{bmatrix} = \begin{bmatrix} V_1 h \\ 0 \end{bmatrix}$,

and V_1 must also be an isometry. Similarly, there exists $V_2: H \to K$
such that

$$V \begin{bmatrix} 0 \\ h \end{bmatrix} = \begin{bmatrix} 0 \\ V_2 h \end{bmatrix} .$$

Consequently,

$$\begin{bmatrix} \phi_1(a) & \phi(b) \\ \phi^*(c) & \phi_2(d) \end{bmatrix} = V^* \begin{bmatrix} \pi(a) & \pi(b) \\ \pi(c) & \pi(d) \end{bmatrix} V = \begin{bmatrix} V_1^*\pi(a)V_1 & V_1^*\pi(b)V_2 \\ V_2^*\pi(c)V_1 & V_2^*\pi(d)V_2 \end{bmatrix} ,$$

which completes the proof of the theorem. □

Unlike the Stinespring representation of a completely positive map,
there are no extra conditions known to impose on the above representation of
a completely bounded map which will make it unique up to unitary
equivalence. Given ϕ , one can always take its minimal, (and hence
unique) Stinespring representation, but unfortunately ϕ is not uniquely
determined by ϕ . It is possible to impose conditions on the above

representation such that it is unique up to conjugation by a (generally
unbounded) closed, densely defined similarity [87].

Theorem 7.5 (Wittstock's Decomposition Theorem). Let A be a
C^*-algebra with unit, and let $\phi: A \to L(H)$ be completely bounded. Then
there exists a completely positive map $\psi: A \to L(H)$ with $\|\psi\|_{cb} \leq \|\phi\|_{cb}$
such that $\psi \pm \mathrm{Re}\ (\phi)$ and $\psi \pm \mathrm{Im}\ (\phi)$ are all completely positive. In
particular, the completely bounded maps are the linear span of the
completely positive maps.

Proof. Let $\phi(a) = V_1^*\pi(a)V_2$ be as in Theorem 7.4 with
$\|V_1\| = \|V_2\| = \|\phi\|_{cb}^{\frac{1}{2}}$, and set $\psi(a) = (V_1^*\pi(a)V_1 + V_2^*\pi(a)V_2)/2$, so that
ψ is completely positive and $\|\psi\|_{cb} = \|\psi(1)\| \leq \|\phi\|_{cb}$. Notice that
$\phi^*(a) = V_2^*\pi(a)V_1$, so that

$$2\psi(a) \pm 2\mathrm{Re}\ \phi(a) = (V_1 \pm V_2)^*\pi(a)(V_1 \pm V_2) ,$$

$$2\psi(a) \pm 2\mathrm{Im}\ \phi(a) = (V_1 \mp iV_2)^*\pi(a)(V_1 \mp iV_2) ,$$

and each of these four maps is completely positive.
    For the last statement, note that

$$2\phi \ = \ ((\psi + \mathrm{Re}\ \phi) - (\psi - \mathrm{Re}\ \phi)) + i((\psi + \mathrm{Im}\ \phi) - (\psi - \mathrm{Im}\ \phi))$$

is a decomposition of ϕ into the span of four completely positive maps. $\square$

7.6 Operator-valued Measures. Let X be a compact Hausdorff space,
let E be a bounded, regular, operator-valued measure on X , and let
$\phi: C(X) \to L(H)$ be the bounded, linear map associated with E by

integration (4.5). We call E. completely bounded when ϕ is completely bounded.

By Wittstock's decomposition theorem, E is completely bounded if and only if it can be expressed as a linear combination of positive operator-valued measures.

When H is 1-dimensional, that is, when E is a bounded, regular, complex-valued measure on X , then the associated bounded, linear map $\phi: C(X) \to \mathbb{C}$ is automatically completely bounded by Proposition 3.7. Wittstock's decomposition becomes the statement that every complex measure is the span of four positive measures. If we let ψ be defined by integration against the total variation measure $|E|$ associated with E , then $\psi \pm \mathrm{Re}\ \phi$ and $\psi \pm \mathrm{Im}\ \phi$ are completely positive maps, and

$$\|\psi\|_{cb} = \|\psi(1)\| = |E|(X) = \|\phi\| = \|\phi\|_{cb} .$$

Thus, we see that in some sense, the ψ of Wittstock's decomposition can be thought of as a total variation or an "absolute value" of the completely bounded map ϕ . This analogy is pursued in [72] and [87].

In a closely related development, Loebl [72] defines a self-adjoint map $\phi: C(X) \to L(H)$ to have <u>finite</u> <u>total</u> <u>variation</u> if

$$\sup \{ \Sigma \|\phi(f_i)\| \| : f_i \text{ is a partition of unity}\}$$

is finite, and proves that every such map decomposes as $\phi = \phi_1 - \phi_2$ with ϕ_1 and ϕ_2 (completely) positive. In particular, this shows that such maps are completely bounded. However, Hadwin [47] has shown that there are completely bounded, self-adjoint maps which are not of finite total variation.

Unfortunately, there is no known analytic characterization of the completely bounded, operator-valued measures. To illustrate the

108

difficulties, let us suppose for simplicity that E is self-adjoint and completely bounded.

From Wittstock's decomposition, we obtain a positive, operator-valued measure F , such that $F(B) \pm E(B) \geq 0$ for all Borel sets B . If E and F were scalar-valued. this inequality would imply that $F(B) \geq |E(B)|$, but for operators, this is far from the case.

In fact, it is possible for E to be completely bounded while

$$\sup \{\|\Sigma \; |E(B_i)|\| : B_i \; \text{disjoint, Borel}\} = +\infty \; .$$

This phenomena was first described by Hadwin [47] and we reproduce his example here.

Let $X = \{x_n\}_{n=1}^{\infty}$ be a countable, compact, Hausdorff space, and let A_n , n = 1, 2, ..., be a sequence of self-adjoint operators on H . If these operators are summable in the weak operator topology, then setting $E(\{x_n\}) = A_n$ defines a self-adjoint. operator-valued measure on X .

Now, let H be a separable Hilbert space with orthonormal basis, $e_1, e_2, \ldots,$ and define A_n via

$$A_n f = 1/n \; (<f,e_1>e_n + <f,e_n>e_1) \; ,$$

so that

$$|A_n| f = 1/n \; (<f,e_1>e_1 + <f,e_n>e_n) \; .$$

It is easily checked that the A_n's are summable in the weak operator topology, but that the $|A_n|$'s are not, since $<|A_n|e_1,e_1>$ is the harmonic series. Thus, setting $F(\{x_n\}) = |A_n|$ will not define an operator-valued measure. However, setting

$$B_n f = 1/n^2 \ <f,e_1>e_1 \ + \ <f,e_n>e_n$$

yields a sequence of positive operators that are summable in the weak
operator topology and satisfy

$$B_n \pm A_n \geq 0 \ .$$

Thus, setting $F(\{x_n\}) = B_n$ yields a positive operator-valued
measure such that $F \pm E$ are positive.

It is easy to see that the linear map $\phi: C(X) \to L(H)$ induced by
E is completely bounded, but not of finite total variation.

Note that the decomposition of $A_n = (B_n + A_n)/2 - (B_n - A_n)/2$ into
the difference of two positive matrices is not the usual decomposition,
$A_n = A_n^+ - A_n^-$, for if it was, then necessarily, we would have
$B_n = A_n^+ + A_n^- = |A_n|$.

In fact, since $\{|A_n|\}$ is not summable, either $\{A_n^+\}$ or $\{A_n^-\}$ must
not be summable. This happens in spite of the fact that we can decompose
each A_n into a difference of positive operators, namely $(B_n \pm A_n)/2$,
with both of these sequences of positive operators summable.

There is another instance in the theory of completely bounded maps
where the usual decomposition of a self-adjoint operator A into A^+ and
A^- is not the "best" decomposition into a difference of positive operators.
This occurs in the theory of Schur products.

7.7 Schur Products Revisited. Let A be in M_n and let
$S_A: M_n \to M_n \cong L(\mathbb{C}^n)$ be given by $S_A(B) = A * B$. Suppose that $\|S_A\|_{cb} \leq 1$,
and using Theorem 7.4, write $S_A(B) = V_1^* \pi(B) V_2$ where $V_i: \mathbb{C}^n \to H$,

110

$i = 1, 2$, are isometries and $\pi: M_n \to L(H)$ is a *-homomorphism. Let $e_1, \ldots, e_n$ be the canonical basis for $\mathbb{C}^n$ and define $2n$ vectors in H via

$$x_j = 1/\sqrt{n} \, \Sigma_{\ell=1}^n \, \pi(E_{\ell,j}) V_2 e_j \, , \quad j = 1, \ldots, n \, ,$$

$$y_i = 1/\sqrt{n} \, \Sigma_{k=1}^n \, \pi(E_{k,i}) V_1 e_i \, , \quad i = 1, \ldots, n \, .$$

We have

$$\| x_j \|^2 = 1/n \, \Sigma_{\ell,m} \, \langle V_2^* \pi(E_{j,\ell} \cdot E_{m,j}) V_2 e_j, \, e_j \rangle$$

$$= \langle V_2^* \pi(E_{j,j}) V_2 e_j, \, e_j \rangle$$

$$\leq 1 \, , \quad j = 1, \ldots, n \, ,$$

and similarly, that

$$y_i^2 \leq 1 \, , \quad i = 1, \ldots, n \, .$$

Let $a_{i,j}$ denote the (i,j)-th entry of A , then

$$\langle x_j, y_i \rangle = 1/n \, \Sigma_{\ell,k} \, \langle V_1^* \pi(E_{i,k} E_{\ell,j}) V_2 e_j, \, e_i \rangle$$

$$= \langle V_1^* \pi(E_{i,j}) V_2 e_j, \, e_i \rangle$$

$$= \langle S_A(E_{i,j}) e_j, \, e_i \rangle$$

$$= a_{i,j} \, .$$

Thus, if $\|S_A\|_{cb} \leq 1$, then there is a Hilbert space and $2n$ vectors in its unit ball such that

$$A = (\langle x_j, y_i \rangle) \, .$$

111

Conversely, if A has such a form, let $P_1 = (<x_j, x_i>)$, $P_2 = (<y_j, y_i>)$, and note that

$$\begin{bmatrix} P_1 & A \\ A & P_2 \end{bmatrix} \geq 0 .$$

By Exercise 7.7, we have that $\|S_A\|_{cb} \leq 1$. Thus, $\|S_A\|_{cb} \leq 1$ if and only if $A = (<x_j, y_i>)$ with $\|x_j\| \leq 1$ and $\|y_i\| \leq 1$, $i,j = 1, \ldots, n$.

It is interesting to see how Wittstock's decomposition theorem relates to these matrices. Suppose that $A = A^*$, so that $S_A = S_A^*$. Since $|A| \pm A \geq 0$, we have that $S_{|A|} \pm S_A \geq 0$, and so $\|S_A\|_{cb} \leq \|S_{|A|}\|_{cb}$. If equality holds in this last inequality, then $S_{|A|}$ could play the role of ψ in the Wittstock decomposition theorem for $\phi = S_A$. However, we shall see (Exercise 7.7vii) that, in general, $\|S_A\|_{cb} < \|S_{|A|}\|_{cb}$. Surprisingly, when $A = A^*$, there always is a positive matrix P with $P \pm A \geq 0$ and $\|S_P\|_{cb} = \|S_A\|_{cb}$ (Exercise 7.7v). Thus, the usual decomposition, $A = \frac{1}{2}(|A| + A) - \frac{1}{2}(|A| - A) = A^+ - A^-$, of a self-adjoint matrix into a difference of positive matrices, while minimal in some senses, is not minimal for the Wittstock decomposition of S_A .

Let M be an operator space, let $\phi: M \to B$, and let ϕ and S be as in Lemma 7.1. It is easy to see that if $\|\phi_k\| \leq 1$, then the proof of Lemma 7.1 shows that ϕ is k-positive. Thus, if $\phi: M \to M_n$ and $\|\phi_{2n}\| \leq 1$, then $\phi: S \to M_2(M_n) = M_{2n}$ will be 2n-positive, and consequently completely positive. But, if ϕ is completely positive, then ϕ must be completely contractive. Hence, $\|\phi_{2n}\| = \|\phi\|_{cb}$ for maps into M_n . Because of the special nature of the subspace S , it turns out that if ϕ if n-positive, then it is completely positive, and consequently,

112

$\|\phi_n\| = \|\phi\|_{cb}$. To obtain this more delicate result, we need a preliminary lemma.

Lemma 7.8. Let A be a C^*-algebra, let $P = (P_{i,j})_{i,j=1}^{2n}$ be a positive element in $M_{2n}(M_2(A))$ where

$$P_{i,j} = \begin{bmatrix} a_{i,j} & b_{i,j} \\ c_{i,j} & d_{i,j} \end{bmatrix} \qquad i,j = 1, \ldots, 2n .$$

Then

$$\left[\begin{bmatrix} a_{i,j} & b_{i,j+n} \\ c_{i+n,j} & d_{i+n,j+n} \end{bmatrix} \right]_{i,j=1}^{n}$$

is positive in $M_n(M_2(A))$.

Proof. Set $A = (a_{i,j})$, $B = (b_{i,j})$, $C = (c_{i,j})$, and $D = (d_{i,j})$ for $i,j = 1, \ldots, 2n$. After the canonical shuffle, P becomes

$\begin{bmatrix} A & B \\ C & D \end{bmatrix}$, and so this latter matrix is positive. Now partition each of

the $2n \times 2n$ matrices into 2×2 matrices consisting of $n \times n$ blocks in the natural fashion. That is,

$$A = \begin{bmatrix} A_{11} & A_{12} \\ A_{21} & A_{22} \end{bmatrix} ,$$

where $A_{11} = (a_{i,j})_{i,j=1}^{n}$, $A_{12} = (a_{i,j+n})_{i,j=1}^{n}$, $A_{21} = (a_{i+n,j})_{i,j=1}^{n}$, and $A_{22} = (a_{i+n,j+n})_{i,j=1}^n$ with the same definitions for B, C, and D.

Thus,

$$
\begin{bmatrix} A & B \\ C & D \end{bmatrix} = \begin{bmatrix} A_{11} & A_{12} & B_{11} & B_{12} \\ A_{21} & A_{22} & B_{21} & B_{22} \\ C_{11} & C_{12} & D_{11} & D_{12} \\ C_{21} & C_{22} & D_{21} & D_{22} \end{bmatrix}
$$

is positive.

A moment's reflection shows that if such a matrix is positive, then the four corners will form a positive matrix, that is,

$$
\begin{bmatrix} A_{11} & B_{12} \\ C_{21} & D_{22} \end{bmatrix}
$$

will be positive.

The result follows by performing the canonical shuffle on this last matrix. □

Proposition 7.9 (Smith). Let M be an operator space and let $\phi: M \to M_n$ be a linear map, then $\|\phi_n\| = \|\phi\|_{cb}$.

Proof. It is sufficient to assume that $\|\phi_n\| = 1$, and prove that $\Phi: S \to M_{2n}$ is completely positive, where ϕ and S are defined as in Lemma 7.1. By Theorem 5.1, to prove ϕ is completely positive, it is sufficient to prove that the associated linear functional $S_\phi: M_{2n}(S) \to \mathbb{C}$ is positive.

114

Thus, let $P = (P_{i,j})_{i,j=1}^{2n}$ be positive in $M_{2n}(S)$, with

$$P_{i,j} = \begin{bmatrix} \lambda_{i,j} & a_{i,j} \\ b_{i,j}^* & \mu_{i,j} \end{bmatrix}, \text{ and calculate}$$

$$
\begin{aligned}
2n \cdot S_\phi(P) &= \Sigma_{i,j=1}^n \phi(P_{i,j})_{(i,j)} + \Sigma_{i,j=1}^n \phi(P_{i,j+n})_{(i,j+n)} \\
&\quad + \Sigma_{i,j=1}^n \phi(P_{i+n,j})_{(i+n,j)} + \Sigma_{i,j=1}^n \phi(P_{i+n,j+n})_{(i+n,j+n)} \\
&= \Sigma_{i=1}^n \lambda_{i,i} + \Sigma_{i,j=1}^n \phi(a_{i,j+n})_{(i,j+n)} \\
&\quad + \Sigma_{i,j=1}^n \phi(b_{i,j})^*_{(i+n,j)} + \Sigma_{i=1}^n \mu_{i+n,i+n} \\
&= \Sigma_{i,j=1}^n \phi(A_{i,j})_{(i,j)} \quad ,
\end{aligned}
$$

where $A_{i,j} = \begin{bmatrix} \lambda_{i,j} & a_{i,j+n} \\ b_{i+n,j}^* & \mu_{i+n,j+n} \end{bmatrix}$. By the above lemma, $A = (A_{i,j})$

is a positive element of $M_n(S)$, and if we set $x = e_1 \oplus \dots \oplus e_n$, then

the last sum above is easily recognized as $<\Phi_n(A)x, x>$. But this last

expression is positive because the assumption that $\|\phi_n\| = 1$ is enough to

guarantee that ϕ is n-positive, as observed in the remarks proceeding

Lemma 7.8. $\qquad\qquad\qquad\qquad\qquad\qquad\qquad\qquad\qquad\qquad\qquad\qquad\qquad$ □

The counterpart of Theorem 3.12 for completely bounded maps is not

true. If $\phi: M_n \to B$, then, in general, $\|\phi_n\| \neq \|\phi\|_{cb}$ and in fact,

there does not exist any integer $m = m(n)$ such that $\|\phi_m\| = \|\phi\|_{cb}$ for all ϕ . This follows from recent work of Haagerup [46].

Combining Proposition 2.9 with Exercise 3.10ii, we see that for maps into M_n , $\|\phi\|_{cb} \leq n\|\phi\|$, which gives another proof of Exercise 3.11.

NOTES

Wittstock obtained his decomposition theorem for completely bounded maps in [123] and proved the extension theorem for completely bounded maps in [124]. His techniques involved the use of his generalization of the Hahn-Banach theorem to set valued mappings into $L(H)$, which in turn rested on some deep results of Connes [29].

Haagerup [45] obtained these same results by exploiting the correspondence between maps from M into M_n and linear functionals on $M_n(M)$, as was done for completely positive maps in section 4. The difficult part of this approach in the completely bounded case is that the norm of the associated linear functional and the cb-norm of the mapping into M_n are not very closely related. Thus, for example, to obtain an extension of a map from M into M_n to A into M_n , where $M \subseteq A$. with the same cb-norm, one must extend the associated linear functional on $M_n(M)$ to $M_n(A)$, but in a very particular fashion. This technical stumbling block was the main obstruction to the theory of completely bounded maps developing simultaneously with the theory of completely positive maps.

The "off-diagonalization" technique first appeared in [84] and was further exploited in [85].

The results in [45], [85], and [124] were done independently.

Proposition 7.9 was obtained by Smith [99].

Haagerup [45] proves the more difficult result that for the Schur product map, $\|S_A\| = \|S_A\|_{cb}$, and obtained the characterization presented in this section of those A's for which S_A is contractive.

The generalization of the "off-diagonalization" technique to C-bimodule maps (Exercise 7.6) is in Suen [108]. Wittstock had proved the decomposition and extension theorems in this more general setting earlier ([123] and [124]).

Exercise 7.8 is proven in Haagerup [46], where a partial converse of the Wittstock decomposition theorem is obtained. Namely, if a von Neumann algebra has the property that the span of the completely positive maps from it into itself is the completely bounded maps, then it is injective. An example is given by Huruya [61] of a C^*-algebra which is not injective, but such that every completely bounded map into it is in the span of the completely positive maps. However, this decomposition does not meet the norm inequality in the statement of the Wittstock decomposition property. It is not known whether every C^*-algebra which has the Wittstock decomposition property, including the norm inequality, is injective. See [103] for more on this topic.

Smith [98] proved that the C^*-algebra $C([0,1])$ has the property that not every completely bounded map from it to itself is in the span of the completely positive maps.

EXERCISES

7.1 Show that $Re\ (\phi_n) = (Re\ \phi)_n$, and that $(\phi^*)_n = (\phi_n)^*$.

7.2 Let $\phi: M \to B$, let H , K be in M_n , and let A be in $M_n(M)$.
Prove that $\phi_n(H \cdot A \cdot K) = H \cdot \phi_n(A) \cdot K$.

7.3 Verify the claim of Theorem 7.4.

7.4 Show that if ϕ is completely bounded, and $\phi(a) = V_1^* \pi(a) V_2$ is the representation of Theorem 7.4 with $\|V_1\| = \|V_2\|$, then setting $\phi_i(a) = V_i^* \pi(a) V_i$ yields the map Φ of Theorem 7.3.

7.5 Prove that the conclusions of Theorems 7.2, 7.3, and 7.5 still hold when the range is changed from $L(H)$ to an arbitrary injective C^*-algebra.

7.6 Let A , B , and C be C^*-algebras with unit, with C contained in both A and B , and $1_A = 1_C$, $1_B = 1_C$. Let $M \subseteq A$ be a subspace such that $c_1 \cdot M \cdot c_2 \subseteq M$ for all c_1 , c_2 in C , and set

$$
S = \left\{ \begin{bmatrix} c_1 & a \\ b^* & c_2 \end{bmatrix} : a, b \in M , \quad c_1, c_2 \in C \right\} .
$$

i) Prove that if $\phi: M \to B$ is a completely contractive C-bimodule map, then $\Phi: S \to M_2(B)$ defined by

$$
\Phi \left(\begin{bmatrix} c_1 & a \\ b^* & c_2 \end{bmatrix} \right) = \begin{bmatrix} c_1 & \phi(a) \\ \phi(b)^* & c_2 \end{bmatrix} ,
$$

is completely positive.

ii) Prove that if B is injective, then the conclusions of Theorems 7.2, 7.3, and 7.5 still hold with the additional requirement that the maps be C-bimodule maps.

7.7 Let $S_A: M_n \to M_n$ be the Schur product map, $S_A(B) = A * B$. Let $A = H + iK$ be the decomposition of A into its real and imaginary parts.

i) Show that $\text{Re}(S_A) = S_H$.

ii) Prove that there exist positive matrices P_1 and P_2 such that

$\phi_i = S_{P_i}$, $i = 1, 2$, satisfy the conclusions of Theorem 6.3

for $\phi = S_A$. [Hint: Recall Exercise 4.4] .

iii) Prove that P_1 and P_2 are such a pair of positive matrices

if and only if $\|S_{P_1}\|_{cb} = \|S_{P_2}\|_{cb} = \|S_A\|_{cb}$ and

$$\begin{bmatrix} P_1 & A \\ A^* & P_2 \end{bmatrix}$$ is positive.

iv) Let $\|S_A\|_{cb} \leq 1$ and write $A = (<x_j, y_i>)$ as in 7.7. Show

that $P_1 = (<y_j, y_i>)$, $P_2 = (<x_j, x_i>)$ are such a pair of

positive matrices.

v) Let $d(B)$ denote the maximum diagonal element of a matrix.

Conclude that

$$\|S_A\|_{cb} = \inf \quad d\begin{bmatrix} P_1 & A \\ A^* & P_2 \end{bmatrix} : \begin{bmatrix} P_1 & A \\ A^* & P_2 \end{bmatrix} \geq 0 \quad .$$

vi) Show that

$$\begin{bmatrix} |A^*| & A \\ A^* & |A| \end{bmatrix} \geq 0 \quad .$$

vii) Give an example where $\|S_A\|_{cb} < \max \{d(|A^*|), d(|A|)\}$, with

$A = A^*$.

7.8 (Haagerup) Let $\phi_1, \phi_2, \phi: A \rightarrow B$. Prove that $\phi: M_2(A) \rightarrow M_2(B)$,

defined by $\phi\begin{bmatrix} a & b \\ c & d \end{bmatrix} = \begin{bmatrix} \phi_1(a) & \phi(b) \\ \phi^*(c) & \phi_2(d) \end{bmatrix}$, is completely positive

if and only if $\psi: A \rightarrow M_2(B)$, defined by $\psi(a) = \begin{bmatrix} \phi_1(a) & \phi(a) \\ \phi^*(a) & \phi_2(a) \end{bmatrix}$,

is completely positive.

7.9 Let $\Phi: M_2(M_n) \to M_2(B)$ be given by $\Phi \begin{bmatrix} a & b \\ c & d \end{bmatrix} = \begin{bmatrix} \phi^+(a) & \phi(b) \\ \phi^*(c) & \phi^-(d) \end{bmatrix}$,

and let $E = (E_{i,j})$ be in $M_n(M_n)$, where $E_{i,j}$ are the canonical matrix units. Prove that Φ is completely positive if and only if

$$\begin{bmatrix} \phi_n^+(E) & \phi_n(E) \\ \phi_n^*(E) & \phi_n^-(E) \end{bmatrix} \quad \text{is positive.}$$

7.10 Let $\phi: M_n \to B$, set $B = (\phi(E_{i,j}))$ in $M_n(B)$, and let $|B^*| = (p_{i,j})$, $|B| = (q_{i,j})$. Define linear maps $\phi_1, \phi_2: M_n \to B$ by

$$\phi_1(E_{i,j}) = p_{i,j}, \quad \phi_2(E_{i,j}) = q_{i,j}.$$

  i) Prove that $\Phi: M_2(M_n) \to M_2(B)$, defined by

$$\Phi\left(\begin{bmatrix} a & b \\ c & d \end{bmatrix}\right) = \begin{bmatrix} \phi_1(a) & \phi(b) \\ \phi^*(c) & \phi_2(d) \end{bmatrix}, \quad \text{is completely positive.}$$

  ii) Prove that $\|\phi\|_{cb}^2 \le \|p_{11} + \dots + p_{nn}\|^2 \cdot \|q_{11} + \dots + q_{nn}\|^2$, and that this estimate is sharp.

  iii) Let $B^*B = (c_{i,j})$. Prove that $\|c_{11} + \dots + c_{nn}\| \le n \cdot \|\phi\|_{cb}^2$, and that this estimate is sharp.

7.11 Prove directly that if $\phi: C(X) \to L(H)$ has finite total variation, then ϕ is completely bounded.

7.12 Let E be a completely bounded, operator-valued measure. Prove that E has finite total 2-variation, i.e., that

$$\sup \{ \Sigma \| E(B_i) \|^2 : B_i \text{ disjoint, Borel} \} < +\infty.$$

7.13 Say E has finite total p-variation if

sup $\{ \Sigma \| E(B_i) |^P \|: B_i$ disjoint, Borel$\} < +\infty$. Does E have finite total 1-variation if and only if its associated map, $\phi: C(X) \to L(H)$, has finite total variation? If E is completely bounded, must E have finite total p-variation for all $p > 1$?

7.14 (Sakai) Let A be a C^*-algebra and let $f: A \to \mathbb{C}$ be a bounded, self-adjoint linear functional with $\| f \| \leq 1$. Show that $f = f_1 - f_2$, with f_1 and f_2 positive linear functionals, and $\| f_1 + f_2 \| = 1$.

8 Completely bounded homomorphisms

In Chapter 6 we saw that if B is a C^*-algebra with unit, and A is a subalgebra of B containing 1_B, then the unital, completely contractive homomorphisms of A into $L(H)$ are precisely the homomorphisms with B-dilations. In this chapter, we prove that the unital homomorphisms of A, which are similar to homomorphisms with B-dilations, are precisely the completely bounded homomorphisms.

If A is a C^*-algebra, then every unital, contractive homomorphism is a positive map and hence a *-homomorphism. Thus, for unital maps of C^*-algebras, the sets of contractive homomorphisms, completely contractive homomorphisms, and *-homomorphisms coincide. In this case, the above result says that a unital homomorphism of a C^*-algebra is similar to a *-homomorphism if and only if it is a completely bounded homomorphism.

Let's begin with a simple observation. Suppose that S is a similarity, ρ is a homomorphism of some operator algebra A, and that $\pi(\cdot) = S^{-1}\rho(\cdot)S$ is a completely contractive homomorphism. Letting S_n denote the direct sum of n copies of S, we have that $S_n^{-1} = (S^{-1})_n$. $\|S_n\| = \|S\|$, and that $\rho_n(\cdot) = S_n\pi_n(\cdot)S_n^{-1}$. Thus, ρ is completely bounded with $\|\rho\|_{cb} \le \|S^{-1}\| \cdot \|S\|$. So any homomorphism which is similar to a completely contractive homomorphism is necessarily completely bounded with

$$\|\rho\|_{cb} \le \inf \{\|S^{-1}\| \cdot \|S\| : S^{-1}\rho(\cdot)S \text{ is completely contractive}\} .$$

The next result proves that not only is the converse of the above

·statement true, but that the above infinium is achieved and gives the cb-norm of ρ .

Theorem 8.1. Let A be an operator algebra and let $\rho: A \rightarrow L(H)$ be a unital, completely bounded homomorphism. Then there exists an invertible operator S , with $\|S\| \cdot \|S^{-1}\| = \|\rho\|_{cb}$ such that $S^{-1}\rho(\cdot)S$ is a completely contractive homomorphism. Moreover,

$$\|\rho\|_{cb} = \inf \{\|R^{-1}\| \cdot \|R\|: R^{-1}\rho(\cdot)R \text{ is completely contractive}\} .$$

Proof. The remarks preceeding the statement of the theorem yield the last equality.

Let A be contained in a C^*-algebra B . By the extension and representation theorems for completely bounded maps, we know that there exists a Hilbert space K , a *-homomorphism $\pi: B \rightarrow L(K)$, and two bounded operators $V_i: H \rightarrow K$, $i = 1, 2$, with $\|\rho\|_{cb} = \|V_1\| \cdot \|V_2\|$, such that $\rho(a) = V_1^*\pi(a)V_2$ for all a in A .

For h in H , we define

$$|h| = \inf \{\|\Sigma \pi(a_i)V_2h_i\|: \Sigma \rho(a_i)h_i = h , a_i \in A , h_i \in H\} ,$$

where the infimum is taken over all finite sums. It is easy to see that $|\cdot|$ defines a pseudonorm on H .

If $h = \Sigma \rho(a_i)h_i$, then $\|h\| = \|\Sigma \rho(a_i)h_i\| = \|\Sigma V_1^*\pi(a_i)V_2h_i\|$ $\leq \|V_1^*\| \cdot \|\Sigma \pi(a_i)V_2h_i\|$, and thus $\|h\| \leq \|V_1^*\| \cdot |h|$. Similarly, the equation $\rho(1)h = h$ yields $|h| \leq \|V_2\| \cdot \|h\|$.

Thus, $|\cdot|$ is equivalent to the original norm on H . Hence $|\cdot|$ is also a norm on H and $(H, |\cdot|)$ is complete.

We claim that $(H, |\cdot|)$ is also a Hilbert space. By the Jordan-von Neumann theorem, it is enough to verify that $|\cdot|$ satisfies the parallelogram law

$$|h + k|^2 + |h - k|^2 = 2(|h|^2 + |k|^2) \;,$$

for all h and k in H.

Let $h = \Sigma \, \rho(a_i)h_i$, $k = \Sigma \, \rho(b_i)k_i$, then $h \pm k = \Sigma\rho(a_i)h_i \pm \Sigma\rho(b_i)k_i$, and so,

$$|h + k|^2 + |h - k|^2 \leq \|\Sigma\pi(a_i)V_2h_i + \Sigma\pi(b_i)V_2k_i\|^2 + \|\Sigma\pi(a_i)V_2h_i - \Sigma\pi(b_i)V_2k_i\|^2$$

$$= 2\|\Sigma \, \pi(a_i)V_2h_i\|^2 + 2\|\Sigma \, \pi(b_i)V_2k_i\|^2 \;.$$

Taking the infimum over all such sums yields

$$|h + k|^2 + |h - k|^2 \leq 2|h|^2 + 2|k|^2 \;.$$

The other inequality follows by substituting $h + k$ and $h - k$ for h and k , respectively. Hence, $(H, |\cdot|)$ is also a Hilbert space.

Let $S: (H, |\cdot|) \to (H, \|\cdot\|)$ be the identity map. Then S is bounded, invertible, $\|S^{-1}\| \cdot \|S\| \leq \|V_1^*\| \cdot \|V_2\| = \|\rho\|_{cb}$, and $S^{-1}\rho(\cdot)S$ is just the homomorphism ρ , but with respect to the $|\cdot|$-norm. Thus, to complete the proof of the theorem, it is sufficient to prove that ρ is completely contractive with respect to this new norm. (To see this, let $U: (H, \|\cdot\|) \to (H, |\cdot|)$ be a unitary and set $R = US$.)

It is not difficult to see that $\rho(\cdot)$ is contractive with respect to the $|\cdot|$-norm, for if $a \in A$, $h \in H$, and $h = \Sigma \, \rho(a_i)h_i$, then

$$|\rho(a)h| = |\Sigma \, \rho(aa_i)h_i| \leq \|\Sigma \, \pi(aa_i)V_2h_i\| \leq \|a\| \cdot \|\Sigma \, \pi(a_i)V_2h_i\| \;,$$

and so, $|\rho(a)h| \leq \|a\| \cdot |h|$.

124

To see that $\rho(\cdot)$ is completely contractive in the $|\cdot|$-norm, fix an integer n, and let $\hat{H} = H \oplus \ldots \oplus H$ (n copies). Let $|\cdot|_n$ denote the Hilbert space norm on $\hat{H}$ induced by $|\cdot|$, that is,

$$|\hat{h}|_n^2 = |h_1|^2 + \ldots + |h_n|^2 ,$$

for $\hat{h} = (h_1, \ldots, h_n)$ in $\hat{H}$. We must prove that if $A = (a_{i,j})$ is in $M_n(A)$, then

$$|\rho_n(A)\hat{h}|_n \leq \|A\| \cdot |\hat{h}_n| .$$

Consider $\hat{H}$ with its old norm,

$$\|\hat{h}\|_n^2 = \|h_1\|^2 + \ldots + \|h_n\|^2 .$$

Since ρ is completely bounded, $\rho_n: M_n(A) \to L(\hat{H}, \|\cdot\|_n)$ is a completely bounded homomorphism, and so by the first part of the proof, we can endow $\hat{H}$ with yet another norm $|\cdot|'$ such that ρ_n is contractive in the $|\cdot|'$-norm. By the first part of the proof, to define $|\cdot|'$, we need a Stinespring representation of ρ_n. To this end, let $\hat{K} = K \oplus \ldots \oplus K$ (n copies), $\hat{V}_i = V_i \oplus \ldots \oplus V_i: \hat{H} \to \hat{K}$, $i = 1, 2$, and $\pi_n: M_n(A) \to L(\hat{K})$, so that $\rho_n(\cdot) = \hat{V}_1^* \pi_n(\cdot) \hat{V}_2$. If we define

$$|\hat{h}|' = \inf \{ \| \Sigma \, \rho_n(A_i)\hat{V}_2 \hat{h}_i \|_n : \Sigma \, \rho_n(A_i)\hat{h}_i = \hat{h} \} ,$$

where the infimum is taken over all finite sums, then $\rho_n(\cdot)$ will be contractive in $|\cdot|'$. The proof of the theorem is now completed by showing that $|\cdot|' = |\cdot|_n$, which we leave to the reader (Exercise 7.1). □

Corollary 8.2 (Haagerup). Let A be a C^*-algebra with unit and let $\rho: A \to L(H)$ be a bounded, unital homomorphism. Then ρ is similar to a

*-homomorphism if and only if ρ is completely bounded. Moreover, if ρ is completely bounded, then there exists a similarity S with $S^{-1}\rho(\cdot)S$ a *-homomorphism and $\|S^{-1}\| \cdot \|S\| = \|\rho\|_{cb}$.

The above result sheds considerable light on a question of Kadison [63]. Kadison asked if every bounded homomorphism of a C^*-algebra is similar to a *-homomorphism. The above result shows that this similarity question is equivalent to determining whether or not bounded homomorphisms are necessarily completely bounded. At the present time the similarity question is still open, although there are a number of deep partial results (see [7], [13], [28] and [44]).

There are several important cases where bounded homomorphisms of a C^*-algebra are completely bounded. These are closely tied to the theory of group representations. Let G be a locally compact, topological group and let $L^\infty(G)$ denote the C^*-algebra of bounded, measurable functions on G. Given g in G, we define the (right) translation operators $R_g: L^\infty(G) \to L^\infty(G)$ by $(R_g f)(g') = f(g'g)$. A state $m: L^\infty(G)^* \to \mathbb{C}$ is called a (right) <u>invariant mean</u> on G if $m(R_g f) = m(f)$ for all g in G . The group G is called <u>amenable</u> if there exists an invariant mean on G .

Note that if G is compact, then it is amenable, since we may define

$$m(f) = \int_G f(g) \, dg ,$$

where dg denotes Haar measure.

Another important class of groups which are amenable are the commutative groups. To see this, note that the adjoint maps, $R_g^*: L^\infty(G)^* \to L^\infty(G)^*$ are weak*-continuous and map states to states. Since the space of states is weak*-compact and convex, and since the maps R_g^*

126

form a commutative group of continuous maps on this space, by the Markov-Kakutani fixed point theorem [37], there will exist a fixed point. This fixed point is easily seen to be an invariant mean.

The importance of the existence of an invariant mean is best illustrated in the following result.

Theorem 8.3 (Dixmier). Let G be an amenable group, and let $\rho: G \to L(H)$ be a strongly continuous homomorphism with $\rho(e) = 1$, such that $\|\rho\| = \sup \{\|\rho(g)\|: g \in G\}$ is finite. Then there exists an invertible S in $L(H)$ with $\|S\| \cdot \|S^{-1}\| \leq \|\rho\|^2$ such that $S^{-1}\rho(g)S$ is a unitary representation of G.

Proof. Let m denote an invariant mean on G, and note that for each pair of vectors x, y in H, the function $f_{x,y}(g) = \langle\rho(g)x, \rho(g)y\rangle$ is a bounded, continuous function on G. Set $\langle x,y\rangle_1 = m(f_{x,y})$, and note that since the map $(x,y) \to f_{x,y}$ is sesquilinear, $\langle\ ,\ \rangle_1$ defines a sesquilinear form on H. Also, since m is a positive linear functional, and $f_{x,x}$ is a positive function, this new sesquilinear form is a semi-definite inner product on H.

We shall show that it is bounded above and below by the original inner product. To see this, set $M = \|\rho\|$ and note that since $\|\rho(g)\| \leq M$ and $\|\rho(g)^{-1}\| \leq M$, we have that

$$1/M^2 \leq \rho(g)^*\rho(g) \leq M^2 .$$

Hence, $1/M^2 \langle x,x\rangle \leq f_{x,x}(g) \leq M^2 \langle x,x\rangle$, and applying m to this inequality yields

$$1/M^2 \ <x,x> \ \leq \ <x,x>_1 \ \leq \ M^2 \ <x,x> \ .$$

Thus, $<x,y>_1$ is an inner product on H .

If we let $(H, |\cdot|_1)$ denote our Hilbert space with the norm induced by this new inner product, then just as in the proof of Theorem 8.1, the map $S: (H, \|\cdot\|) \to (H, |\cdot|_1)$ is bounded and invertible, with
$$\|S\| \cdot \|S^{-1}\| \leq M^2 = \|\rho\|^2 \ .$$

Finally, note that with respect to this new inner product,

$$<\rho(g)x, \ \rho(g)y>_1 \ = \ m \ (R_g f_{x,y}) \ = \ m \ (f_{x,y}) \ = \ <x,y>_1 \ .$$

and so $\rho(g)$ is a unitary on $(H, |\cdot|_1)$. ☐

Corollary 8.4 (Sz.-Nagy). Let T be an invertible operator on a Hilbert space such that $\|T^n\| \leq M$ for all integers n . Then there exists an invertible operator S , with $\|S^{-1}\| \cdot \|S\| \leq M^2$, such that $S^{-1}TS$ is a unitary operator.

Proof. Let Z denote the group of integers, define $\rho(n) = T^n$, and apply Theorem 8.3. ☐

In a number of cases, similarity results for homomorphisms of C^*-algebras can be obtained from Theorem 8.3 by restricting attention to the group of unitaries in the C^*-algebra. The following result demonstrates this principle.

Lemma 8.5. Let A be a C^*-algebra. Then every element in A is a linear combination of at most four unitaries.

128

Proof. Let $h = h^*$ be a self-adjoint element of A, with $\|h\| \leq 1$.
Then $u = h + i\sqrt{1 - h^2}$ is easily seen to be unitary, and $h = u + u^*$.
This shows that every self-adjoint element is in the span of two unitaries.
Using the Cartesian decomposition, we obtain the result.   □

Lemma 8.6. Let A and B be C^*-algebras and let $\rho: A \rightarrow B$ be a
homomorphism with $\rho(1) = 1$. If ρ maps unitaries to unitaries, then ρ
is a *-homomorphism.

Proof. Let u be unitary in A, then $\rho(u^*) = \rho(u^{-1}) = \rho(u)^{-1} = \rho(u)^*$,
and so ρ is self-adjoint on the unitary elements of A. By Lemma 8.5,
ρ is self-adjoint on A.   □

Theorem 8.7. Let A be a commutative, unital C^*-algebra. If
$\rho: A \rightarrow L(H)$ is a bounded homomorphism, then ρ is completely bounded and
$\|\rho\|_{cb} \leq \|\rho\|^2$.

Proof. Let G denote the group of unitary elements of A. By
Theorem 8.3, there is a similarity S, with $\|S\| \cdot \|S^{-1}\| \leq \|\rho\|^2$ such that
$S^{-1}\rho(u)S$ is unitary for all u in G. By Lemma 8.6, $S^{-1}\rho(\cdot)S$ is a
*-homomorphism.   □

Corollary 8.8. Let T be an operator on a Hilbert space. Then T
is similar to a self-adjoint operator if and only if for some interval,
$[a,b]$, there is a constant k such that

$$\|p(T)\| \leq K \cdot \sup \{|p(t)|: t \in [a,b]\}$$

129

for all polynomials p with real coefficients.

An affirmative answer to Kadison's similarity question for C^*-algebras

is known to imply an affirmative answer to two other questions concerning

derivations and invariant operator ranges. Consequently. these problems

have an analogous status, they have affirmative answers if and only if

certain bounded maps are completely bounded, but at the present time. both

questions are still open. We discuss the derivation question in this

section. For a discussion of invariant operator ranges and their connections

with completely bounded maps, see [84].

Let $A \subseteq L(H)$ be a C^*-algebra. A linear map $\delta: A \rightarrow L(H)$ is called

a derivation if $\delta(AB) = A\delta(B) + \delta(A)B$. It is the case that every deriva-

tion is automatically bounded [91]. If $X \in L(H)$, then setting

$\delta(A) = AX - XA$ defines a derivation and such a derivation is called inner.

The derivation question asks whether every derivation is necessarily inner.

Given a derivation δ , if we define $\rho: A \rightarrow L(H \oplus H)$ by

$$\rho(A) = \begin{bmatrix} A & \delta(A) \\ 0 & A \end{bmatrix},$$

then ρ will be a homomorphism of A . In fact, it is not difficult to see

that ρ is a homomorphism if and only if δ is a derivation.

Proposition 8.9. The derivation δ is inner if and only if ρ is

similar to a *-homomorphism.

Proof. First. suppose that δ is inner, so that $\delta(A) = AX - XA$.

130

Setting

$$S = \begin{bmatrix} 1 & X \\ 0 & 1 \end{bmatrix}, \qquad S^{-1} = \begin{bmatrix} 1 & -X \\ 0 & 1 \end{bmatrix},$$

we have that

$$S\rho(A)S^{-1} = \begin{bmatrix} A & 0 \\ 0 & A \end{bmatrix},$$

which is a *-homomorphism.

Conversely, if S is a similarity such that $\pi(A) = S^{-1}\rho(A)S$ is a *-homomorphism, then set $X = SS^*$. We have that

$$\rho(A)X = S\pi(A)S^* = (S\pi(A^*)S^*)^* = (\rho(A^*)SS^*)^* = X\rho(A^*)^*.$$

Writing

$$X = \begin{bmatrix} X_{11} & X_{12} \\ X_{12}^* & X_{22} \end{bmatrix},$$

the above equation becomes

$$\begin{bmatrix} AX_{11} + \delta(A)X_{12}^* & AX_{12} + \delta(A)X_{22} \\ AX_{12}^* & AX_{22} \end{bmatrix} = \begin{bmatrix} X_{11}A + X_{12}\delta(A^*)^* & X_{12}A \\ X_{12}^*A + X_{22}\delta(A^*)^* & X_{22}A \end{bmatrix}.$$

But since X is positive and invertible, X_{22} must also be positive and invertible. Equating the (1,2)-entries of the above operator matrices yields

$$\delta(A) = \delta(A)X_{22}X_{22}^{-1} = (X_{12}A - AX_{12})X_{22}^{-1} = X_{12}X_{22}^{-1}A - AX_{12}X_{22}^{-1},$$

since X_{22} commutes with A .   □

Corollary 8.10 (Christensen). Let $A \subseteq L(H)$ be a unital C^*-algebra
and let $\delta : A \to L(H)$ be a derivation. Then δ is inner if and only if
δ is completely bounded.

Proof. By using the canonical shuffle, it is easy to see that ρ is
completely bounded if and only if δ is completely bounded. But by
Proposition 8.3 and Corollary 8.2, δ is inner if and only if ρ is
completely bounded.   □

Turning our attention to some non-selfadjoint algebras. Theorem 8.1
can be used to give a characterization of the operators that are similar to
a contraction. Let $P(\mathbf{D})$ denote the algebra of polynomials with the norm
it inherits as a subalgebra of $C(\Pi)$.

Theorem 8.11. Let $T \in L(H)$. Then T is similar to a contraction
if and only if the homomorphism $\rho : P(\mathbf{D}) \to L(H)$ defined by $\rho(p) = p(T)$ is
completely bounded. Moreover, if this is the case, then

$$\| \rho \|_{cb} = \inf \{ \| S \| \cdot \| S^{-1} \| : \| S^{-1} TS \| \leq 1 \} ,$$

and the infimum is attained.

Proof. If ρ is completely bounded, then there is a similarity S ,
with $\| S^{-1} \| \cdot \| S \| = \| \rho \|_{cb}$, such that $S^{-1} \rho(\cdot) S$ is completely contractive.
Thus,

132

$$\|S^{-1}TS\| = \|S^{-1}\rho(z)S\| \le \|z\| = 1 .$$

where z is the coordinate function.

Conversely, if $R = S^{-1}TS$ is a contraction, then $\theta: P(\mathbb{D}) \to L(H)$, given by $\theta(p) = p(R)$, is a contractive homomorphism by von Neumann's inequality, and is completely contractive by Sz.-Nagy's dilation theorem. But $\rho(\cdot) = S\theta(\cdot)S^{-1}$, and so ρ is completely bounded with

$$\|\rho\|_{cb} \le \|S\| \cdot \|S^{-1}\| .$$

The statement about the infimum follows from the corresponding statement in Theorem 8.1. □

An operator for which $\rho(p) = p(T)$ is a bounded homomorphism of $P(\mathbb{D})$ is called a polynomially bounded operator, and it is quite natural to refer to the operators for which this ρ is completely bounded as completely polynomially bounded. Halmos [49] has asked if every polynomially bounded operator is similar to a contraction. The above result shows that this question is equivalent to determining whether or not certain bounded homomorphisms are completely bounded. This question is also still open.

The above theory sheds some light on a theory closely related to the theory of spectral sets. Let X be a compact set in $\mathbb{C}$, $R(X)$ the algebra of quotients of polynomials, and let T be in $L(H)$ with $\sigma(T)$ contained in X. If the homomorphism $\rho: R(X) \to L(H)$ defined by $\rho(r) = r(T)$ is bounded with $\|\rho\| \le K$, then X is called a K-spectral set for T. If ρ is completely bounded with $\|\rho\|_{cb} \le K$, we shall call X a complete K-spectral set for T. Recall that X is a complete spectral set if and only if T has a normal ∂X-dilation. It is immediate that:

Corollary 8.12. A set X is a complete K-spectral set for T' if and only if T is similar to an operator for which X is a complete spectral set.

Moreover, there exists such a similarity satisfying $\|S^{-1}\| \cdot \|S\| \leq K$.

In fact, we have that for ρ as above,

$\|\rho\|_{cb}$ = inf $\{\|S^{-1}\| \cdot \|S\| : X$ is a complete spectral set for $S^{-1}TS\}$, and the infimum is attained.

The complete picture of how these concepts are related is still very open. As discussed earlier, it is still not known if a set is a spectral set for an operator, whether or not it must be a complete spectral set for the operator. It is also unknown if a set is a K'-spectral set for an operator, whether or not it must be a complete K-spectral set for the operator, for some K . Halmos' question is the special case of this question where X = $\mathbb{D}^-$.

One fact that is known is that K ≠ K' , that is, examples are known of operators for which $\|\rho\| \neq \|\rho\|_{cb}$, where ρ is the homomorphism of R(X) . Holbrook [58] gives an example of a matrix T , which is similar to a contraction such that

$$\rho \quad < \quad \inf \{\|S^{-1}\| \cdot \|S\| : \|S^{-1}TS\| \leq 1\} ,$$

where ρ is the induced homomorphism of P($\mathbb{D}$) . The theory of completely bounded maps at least gives a basis for explaining such phenomena.

Maps which are defined by some type of analytic functional calculus are often completely bounded. Recall the Riesz functional calculus. If G is an open set containing $\sigma(T)$ and ∂G consists of a finite number of simple, closed, rectifiable curves, then for any function in R(G) , we have

134

that

$$r(T) = 1/2\pi i \int_{\partial G} r(z) \quad (T - zI)^{-1} \, dz \ .$$

This functional calculus can be used to derive what is referred to as the generalized Rota model of an operator.

Theorem 8.13 (Herrero-Voiculescu). Let T be in $L(H)$, and let G be an open set in the complex plane such that $\sigma(T)$ is contained in G and such that ∂G consists of a finite number of simple, closed, rectifiable curves. Then T is similar to an operator that has a normal ∂G-dilation. Moreover, the similarity may be chosen to satisfy

$$\| S^{-1} \| \cdot \| S \| \ \leq \ 1/2\pi \int_{\partial G} \| (T - zI)^{-1} \| \ |dz| \ .$$

Proof. By Corollary 8.12, we need only prove that G^- is a complete K-spectral set for T with

$$K = 1/2\pi \int_{\partial G} \| (T - zI)^{-1} \| \ |dz| \ .$$

Let $\rho \colon R(G^-) \to L(H)$ be the homomorphism defined by $\rho(r) = r(T)$. We must prove that $\| \rho \|_{cb} \leq K$.

By the Riesz functional calculus,

$$\| \rho(r) \| \ \leq \ 1/2\pi \int_{\partial G} \| r(z) \cdot (T - zI)^{-1} \| \ |dz| \ \leq \ K \cdot \| r \| \ ,$$

so $\| \rho \| \leq K$. Let $\hat{T}$ denote the direct sum of n copies of T and note that $\| (\hat{T} - z\hat{I})^{-1} \| = \| (T - zI)^{-1} \|$. Thus, for $(r_{i,j})$ in $M_n(R(G^-))$, we have that

$$\rho_n((r_{i,j})) = 1/2\pi i \int_{\partial G} (r_{i,j}(z) \cdot (T - zI)^{-1}) \, dz$$

$$= 1/2\pi i \int_{\partial G} (r_{i,j}(z)) \cdot (\hat{T} - z\hat{I})^{-1} \, dz \,,$$

and so by the same inequality as above, $\|\rho_n\| \le K$, which proves that G^- is a complete K-spectral set for T . □

Corollary 8.14 (Rota). Let T be in $L(H)$ with $\sigma(T)$ contained in the open unit disk. Then T is similar to a contraction. Moreover, a similarity may be chosen satisfying

$$\|S^{-1}\| \cdot \|S\| \le 1/2\pi \int_\Pi \|(T - zI)^{-1}\| \, |dz| \,.$$

Proof. By Corollary 8.12, T is similar to an operator with a unitary dilation, and that operator is necessarily a contraction. □

The usual proof of Rota's theorem gives much more particular information on the unitary dilation than this proof does. In fact, the unitary can always be taken to be the bilateral shift (see, for example, [39]). One difference is that the above proof can give quite different estimates on

$$\|S^{-1}\| \cdot \|S\| \,.$$

Recall the spectral radius formula,

$$\sup \{|z|: z \in \sigma(T)\} = \lim_n \|T^n\|^{1/n} \,.$$

So that if $\sigma(T)$ is contained in $\mathbb{D}$, then

136

$$(T - e^{i\theta}I)^{-1} = e^{-i\theta} \, \Sigma_n \, (e^{i\theta}T)^n \, ,$$

and the latter series converges by the root test. Thus, we see that the estimate on $\|S^{-1}\| \cdot \|S\|$ obtained in Corollary 8.13 satisfies

$$K \leq \Sigma_n \, \|T^n\| \, .$$

A careful reading of the proof of Rota's theorem in [39, p.21] leads to an upper bound on $\|S\| \cdot \|S^{-1}\|$ of $\Sigma_n \, \|T^n\|^2$.

NOTES

Kadison's study of homomorphisms of C^*-algebras [63] was motivated in part by Dixmier's study of when representations of groups are similar to unitary representations [33]. Of this circle of questions, only Dixmier's is known to have a negative answer, which was obtained by Kunze-Stein [67]. The original proof of Sz.-Nagy's result characterizing operators that are similar to unitaries (Corollary 8.8), contains the elements of Dixmier's result on amenable groups (Theorem 8.3). This proof is outlined in the exercises (Exercise 8.7) and can be found in [109].

Hadwin [47] proved that a unital homomorphism of a C^*-algebra into $L(H)$ is similar to a *-homomorphism if and only if the homomorphism is in the span of the completely positive maps from the algebra to $L(H)$, and conjectured that the span of the completely positive maps was the completely bounded maps. At the same time, Wittstock [121] proved his decomposition theorem which verifies this conjecture.

Simultaneously, Haagerup [44] proved directly that a unital homomorphism of a C^*-algebra into $L(H)$ is similar to a *-homomorphism if and only if it is completely bounded, and moreover, that there exists a similarity S such that $\|S^{-1}\| \cdot \|S\|$ is equal to the cb-norm of the homomorphism. As yet, this equality has not been obtained by combining the techniques of Hadwin and Wittstock. Haagerup applied his result to obtain the analogous characterization of inner derivations (Corollary 8.10). The result on inner derivations had been obtained earlier by Christensen [27].

In [83], the extension theorem for completely bounded maps into $L(H)$ and the generalization of Stinespring's representation theorem were proven in order to extend the techniques of Hadwin/Wittstock to operator algebras. It was proven that a unital homomorphism of an operator algebra into $L(H)$ is similar to a completely contractive homomorphism if and only if it is completely bounded. The prime motivation was to prove that an operator is similar to a contraction if and only if it is completely polynomially bounded (Theorem 8.11), which had been conjectured by Arveson. The extension theorem had been obtained earlier by Wittstock [122].

The fact that the similarity could be chosen such that $\|S\| \cdot \|S^{-1}\|$ is equal to the cb-norm of the homomorphism for a general operator algebra (Theorem 8.1) was obtained later [86]. The key new technique can be found in a paper of Holbrook [56].

The problem of characterizing the operators that are similar to contractions has been considered by Sz.-Nagy. After obtaining his characterization of the operators that are similar to unitaries, he conjectured that an operator T was similar to a contraction if and only if $\|T^n\|$ was uniformly bounded for all positive integers n. Such an operator is called <u>power bounded</u>. Foguel [40] (see also, Halmos [48]) gave an example of a power bounded operator that is not similar to a

138

contraction, and Halmos conjectured that the right condition was polynomially bounded.

EXERCISES

8.1 Verify the claim of Theorem 8.1, that $|\cdot|' = |\cdot|_n$.

8.2 Let A be an algebra and let $\rho: A \to L(H)$ be a homomorphism. If $\rho(A)H$ is dense in H , then ρ is called _non-degenerate_. Show that if A is unital, then ρ is non-degenerate if and only if $\rho(1) = 1$.

8.3 Let P be in $L(H)$ such that $P^2 = P$.

   i) Show that relative to some decomposition of H ,

$$ P = \begin{bmatrix} 1 & X \\ 0 & 0 \end{bmatrix} . $$

   ii) Show that if

$$ S = \begin{bmatrix} 1 & X \\ 0 & 1 \end{bmatrix} , $$

   then SPS^{-1} is an orthogonal projection.

   iii) Define $\rho: \mathbb{C} \oplus \mathbb{C} \to L(H)$ via $\rho(\lambda_1, \lambda_2) = \lambda_1 P + \lambda_2(1 - P)$, and show that ρ is a completely bounded homomorphism, but that in general, $\|\rho\|_{cb} < \|S^{-1}\| \cdot \|S\|$.

   iv) Find $\|\rho\|_{cb}$.

8.4 Let B be a C^*-algebra with unit, let $A \subseteq B$ be a subalgebra which does not contain the identity of B , and let $\rho: A \to L(H)$ be a homomorphism. Set $A_1 = \{a + \lambda 1: a \in A , \lambda \in \mathbb{C}\}$ and define

139

$\rho_1 : A_1 \rightarrow L(H)$ by $\rho_1(a + \lambda 1) = \rho(a) + \lambda \cdot 1_H$. Prove that ρ_1 is a completely bounded homomorphism if and only if ρ is a completely bounded homomorphism.

8.5 Let A be a unital operator algebra, $\rho : A \rightarrow L(H)$ a bounded homomorphism, and $\rho(1) = 1$.

  i) Show that if there exists x_1 in H such that $\rho(A)x_1 = H$, then ρ is completely bounded.

  ii) Show that if there exists $x_1, \ldots, x_n$ in H such that $\rho(A)x_1 + \ldots + \rho(A)x_n = H$, then ρ is completely bounded.

8.6 Prove that if T is a $n \times n$ matrix and X is a K-spectral set for T , then X is a complete (nK)-spectral set for T . Characterize those matrices for which $\mathbb{D}^-$ is a K-spectral set for some K in terms of their Jordan form.

8.7 (Sz.-Nagy) Let T be an invertible operator on H such that $\|T^n\| \leq M$ for all integers n , and let glim be a Banach generalized limit [12].

  i) Show that $<x,y>_1 = \text{glim} <T^n x, T^n y>$ defines a new inner product on H and that $1/M^2 <x,x> \leq <x,x>_1 \leq M^2 <x,x>$.

  ii) Show that T is a unitary transformation on $(H, < \; , \; >_1)$.

  iii) Prove that there exists a similarity S on H , with $\|S^{-1}\| \cdot \|S\| \leq M^2$, such that $S^{-1}TS$ is a unitary.

8.8 Prove that a finite direct sum of operators $T_1 \oplus \ldots \oplus T_n$ is similar to a contraction if and only if each operator in the direct sum is similar to a contraction. Moreover, prove that

140

$$\inf \ \{\|S\| \cdot \|S^{-1}\| \ : \ \|S^{-1}(T_1 \oplus \ldots \oplus T_n)S\| \le 1\}$$

is achieved by an operator S that is itself a direct sum.

8.9 Let T_1, T_2, T_3 be the operators of Parrott's example 6.9, and let ρ be the contractive homomorphism of $P(\mathbb{D}^3)$ defined by $\rho(z_i) = T_i$. Prove that ρ is completely bounded.

8.10 (Sz.-Nagy-Foias) An operator T in $L(H)$ is said to belong to class C_ρ if there exists a Hilbert space K containing H, and a unitary U on K, such that

$$T^n \ = \ \rho P_H U^n |_H \ ,$$

for all positive integers n. Prove that such a T is completely polynomially bounded and that there exists an invertible operator S such that $S^{-1}TS$ is a contraction with $\|S\| \cdot \|S^{-1}\| \le 2\rho - 1$ when $\rho \ge 1$.

8.11 Let X be a compact set in the complex plane such that $R(X)$ is dense in $C(\partial X)$. Prove that X is a K-spectral set for some operator T if and only if T is similar to a normal operator and $\sigma(T)$ is contained in ∂X. Show that the similarity can always be chosen such that $\|S\| \cdot \|S^{-1}\| \le K^2$.

8.12 Let T be an operator with real spectrum. Prove that T is similar to a self-adjoint operator if and only if the Cayley transform of T, $C = (T + i)(T - i)^{-1}$, has the property that for some constant M, $\|C^n\| \le M$ for all integers n.

8.13 Let P be a family of (not necessarily self-adjoint) commuting

projections on a Hilbert space H satisfying:

a) $P \in P$, then $(1 - P) \in P$.

b) P, $Q \in P$, then $PQ \in P$.

c) P, $Q \in P$, $PQ = 0$, then $P + Q \in P$.

d) $\|P\| = \sup \{\|P\| : P \in P\}$ is finite.

i) For P, $Q \in P$, set $P \triangle Q = 1 - P - Q + 2PQ$ and show that

this is an element of P. We call $P \triangle Q$ the <u>symmetric</u>

<u>difference</u> of P and Q.

ii) Prove that $(P, \triangle)$ is a commutative group.

iii) Prove that $\rho(P) = 2P - 1$ defines a representation of this

group and that $\|\rho\| \leq 2\|P\|$.

iv) Prove that there exists a similarity S with

$\|S\| \cdot \|S^{-1}\| \leq 4\|P\|^2$ such that $S^{-1}PS$ is a self-adjoint

projection for all P in P.

v) Use iv) to give an alternative proof of the fact that every

homomorphism of a commutative, unital C^*-algebra is similar

to a *-homomorphism. What estimate can you obtain on

$\|S\| \cdot \|S^{-1}\|$?

8.14 Let A be a finite dimensional C^*-algebra and let $\rho: A \to L(H)$ be

a homomorphism, $\rho(1) = 1$. Show that ρ is completely bounded

and that $\|\rho\|_{cb} \leq \|\rho\|^2$.

9 Applications to K-spectral sets

In this chapter we apply the results of Chapter 8 to the study of multiply-connected K-spectral sets. If a compact subset X of the complex plane is simply connected with boundary a Jordan curve, then one can define an analytic homomorphism f from X to the closed unit disk. In this case it is easily seen that X is a spectral set for an operator T if and only if $f(T)$ is a contraction. For this reason, criteria for operators to have a simply-connected set as a spectral set are fairly readily available. On the other hand, we shall see that criteria for a non-simply connected set to be a spectral set for an operator are considerably subtler. However, we shall see that for many finitely connected sets, there is a simple criterion for the set to be a K-spectral set for an operator. To develop this criterion, we first need to extend the definition of spectral sets to include closed, proper subsets of the complex plane.

To motivate this extended definition, suppose that T is an invertible operator, $S = T^{-1}$, and $\|S\| \leq R$, so that the closed disc of radius R is a spectral set for S. The fact that this disc is a spectral set for S only tells us that $\|S\| \leq R$ and loses the information that S was invertible. The statement that will capture both pieces of information about T, when appropriately defined, is that the complement of the open disc of radius R^{-1} is spectral for T.

Let X be a closed, proper subset of $\mathbb{C}$, and let $\hat{X}$ denote the closure of X, when we regard X as a subset of the Riemann sphere.

That is, $\hat{X} = X$, when X is compact, and otherwise $\hat{X}$ is X together
with the point at infinity. We let R(X) denote the quotients of
polynomials with poles off $\hat{X}$, that is, the bounded, rational functions
on X with a limit at infinity. We regard R(X) as a subalgebra of
$C(\partial \hat{X})$, which defines norms on $M_n(R(X))$.

If X is a closed, proper subset of $\mathbb{C}$, and T $\in$ L(H) with
$\sigma(T) \subseteq X$, then there is still a functional calculus, i.e., a homomorphism
$\rho: R(X) \to L(H)$, given by $\rho(p/q) = p(T)q(T)^{-1}$. We say that X is a
(complete) K-spectral set for T , provided that $\|\rho\| \leq K$ (respectively,
$\|\rho\|_{cb} \leq K$), and use the term (complete) spectral when K = 1 .

Let ψ be a linear fractional transformation regarded as a map from
the sphere to the sphere, let X be a closed, proper subset of $\mathbb{C}$, and
suppose that the pole of ψ lies off X , so that $\psi(\hat{X}) = \psi(X)^- = Y$ is
a compact set in $\mathbb{C}$. If f is in R(Y) , then f $\circ$ ψ defines an element
of R(X) , and the map $\psi^*: R(Y) \to R(X)$, given by $\psi^*(f) = f \circ \psi$,
defines a completely isometric algebra isomorphism between these algebras.
The following results are immediate.

Proposition 9.1. Let X be a closed, proper subset of $\mathbb{C}$, and let
ψ be a linear fractional transformation with pole off X . Then X is
a (complete) K-spectral set for some operator T if and only if $\psi(X)^-$
is a (complete) K-spectral set for $\psi(T)$.

Proposition 9.2. Let T be an operator. Then T is invertible
with $\|T^{-1}\| \leq R$ if and only if $\{z: |z| \geq R^{-1}\}$ is a complete spectral
set for T .

144

It is now quite easy to illustrate one of the subtleties involved in the study of finitely connected spectral sets that is eliminated by the study of K-spectral sets. Let $X = \{z: R^{-1} \leq |z| \leq R\}$, $R > 1$, be a spectral set for some operator T, so that $\|T\| \leq R$ and $\|T^{-1}\| \leq R$. This last statement is equivalent to the statement that $X_1 = \{z: |z| \leq R\}$ and $X_2 = \{z: |z| \geq R^{-1}\}$ are both spectral sets for T. Since $X = X_1 \cap X_2$, it is natural to ask if X_1 and X_2 are spectral sets for T, then is X a spectral set for T? The answer, as we shall show in a moment, is no.

On the other hand, if X_1 and X_2 are, respectively, (complete) K_1-spectral and (complete) K_2-spectral sets for T, then we shall show that X is always a (complete) K-spectral set for T, for some K (Theorem 9.5).

To construct an example where X is not spectral for T, let

$$T = \begin{bmatrix} 1 & t \\ 0 & 1 \end{bmatrix}, \quad T^{-1} = \begin{bmatrix} 1 & -t \\ 0 & 1 \end{bmatrix}.$$

It is not difficult to calculate that, if $t = R - R^{-1}$, then $\|T\| = \|T^{-1}\| = R$. Thus, X_1 and X_2 are both spectral sets for T. However, $|z - z^{-1}| \leq R + R^{-1}$ on X, while $\|T - T^{-1}\| = 2t = 2(R - R^{-1}) > R + R^{-1}$, for $R > \sqrt{3}$. Thus, X is not a spectral set for T, when $R > \sqrt{3}$.

To show that the above annulus is always a K-spectral set requires the introduction of the concept of decomposability. Let X be a closed, proper subset of $\mathbb{C}$. We call a collection $\{X_i\}$ of closed, proper subsets of $\mathbb{C}$, a <u>decomposition</u> of X provided that $X = \cap X_i$ and every $f \in R(X)$ can be written as a uniformly convergent series $f = \Sigma_i f_i$, where each f_i is in $R(X_i)$ and $\Sigma_i \|f_i\| \leq K\|f\|$, where K is independent of f. The least value of K satisfying the above inequality we call the decomposition constant, relative to the decomposition $\{X_i\}$. If there is

a constant K such that the above inequality holds for all f in $M_n(R(X))$,

and all n , then we say that $\{X_i\}$ is a complete decomposition of X

and call the least such K the complete decomposition constant.

Of course every set has a trivial decomposition, namely itself.

However, we shall see that many sets have more interesting decompositions.

Before proceeding, we point out the relevance of the above definitions.

Proposition 9.3. Let X be a closed, proper set in $\mathbb{C}$ with (complete)

decomposition $\{X_i\}$ and (complete) decomposition constant K . If T is

an operator such that each X_i is a (complete) K_i-spectral set for T

and L = $\sup_i K_i$ is finite, then X is a (complete) KL-spectral set for T .

The algebra R(X) is called a Dirichlet algebra if $R(X) + \overline{R(X)}$ is

dense in $C(\partial X)$. We shall call a set X for which R(X) is a Dirichlet

algebra a D-set. This concept should not be confused with the concept of

a Dirichlet set. A set X is called a Dirichlet set if every continuous

function on ∂X has a harmonic extension to the interior of X . For

example, an annulus is a Dirichlet set which is not a D-set.

Let X be a compact set in $\mathbb{C}$ whose boundary consists of n + 1

disjoint, rectifiable, simple, closed curves (i.e., Jordan curves). Such

a set will be called a nice n-holed set. If X is a nice n-holed set, let

$\{U_i\}_{i=0}^n$ denote the open components of $\mathbb{C}\setminus X$, with U_o the unbounded

component, and let $X_i = \mathbb{C}\setminus U_i$, so that X_i , i = 0, ..., n , is a closed

set with X_o compact. Note that X = $X_o \cap X_1 \cap \ldots \cap X_n$. We call

$\{X_i\}_{i=0}^n$ the canonical decomposition of X .

146

· Proposition 9.4. Let X be a nice n-holed set. Then the canonical decomposition of X is a complete decomposition of X.

Proof. Let $\{X_j\}_{j=0}^n$ be the canonical decomposition of X and let $\Gamma_j = \partial X_j$, with Γ_o oriented counterclockwise, and Γ_j, $j = 1, \ldots, n$, oriented clockwise. If $f \in M_k(R(X))$, then for $z \notin \Gamma_j$, set

$$f_j(z) = 1/2\pi i \int_{\Gamma_j} f(w) (w - z)^{-1} dw .$$

Since for z in the interior of X, $f(z) = f_o(z) + \ldots + f_n(z)$, it is not difficult to see that each $f_j(z)$ extends to define a function in $M_k(R(X_j))$, which we still denote by $f_j(z)$, and that $f(z) = f_o(z) + \ldots + f_n(z)$ for all $z \in X$.

For $i \neq j$, let $d_{i,j}$ denote the minimum distance between Γ_i and Γ_j, and let ℓ_i denote the length of Γ_i times $(2\pi)^{-1}$. For $i \neq j$, and $z \in \Gamma_j$, we have that $|f_i(z)| \leq \ell_i d_{i,j}^{-1} \|f\|$. Thus, for $z \in \Gamma_i$,

$$|f_i(z)| = |f(z) - \Sigma_{j \neq i} f_j(z)| \leq \|f\| + \Sigma_{j \neq i} \ell_j d_{i,j}^{-1} \|f\| ,$$

and so we have that $\|f_i\| \leq c_i \|f\|$, where c_i is a constant independent of k.

Hence, $\Sigma_i \|f_i\| \leq (c_o + \ldots + c_n) \|f\|$ and we have that the canonical decomposition is a complete decomposition. □

Theorem 9.5. Let X be a nice n-holed set with canonical decomposition $\{X_i\}_{i=0}^n$, and let $T \in L(H)$. The following are equivalent:

i) X is a (complete) K-spectral set for T, for some K.

ii) Each X_i is a (complete) K_i-spectral set for T.

147

Proof. A straightforward application of Propositions 9.3 and 9.4. □

In Chapter 8 we discussed the question of whether the hypothesis that X is a K-spectral set for T implies that X is a complete K-spectral set for T for some K' . By Theorem 9.5, we see that this question has an affirmative answer for nice n-holed sets if and only if it has an affirmative answer for sets of the form appearing in the canonical decomposition. That is, for simply connected sets whose boundary is a simple, rectifiable, closed curve. By the Riemann mapping theorem, these latter sets are all conformally equivalent to the closed unit disk. Thus, if K-spectral implies complete K'-spectral for the closed unit disk, then the same is true for all nice n-holed sets. If an example exists of an operator for which some nice n-holed set is a K-spectral set, but not a complete K'-spectral set, then such an example exists for the disk.

There is another more geometric version of Theorem 9.5.

Corollary 9.6. Let X be a nice n-holed set, with canonical decomposition $\{X_i\}_{i=0}^{n}$, and let $T \in L(H)$. The following are equivalent:

i) There exists an invertible operator S such that $S^{-1}TS$ has a normal ∂X-dilation.

ii) There exist invertible operators $\{S_i\}_{i=0}^{n}$ such that $S_i^{-1}TS_i$ has a normal ∂X_i-dilation.

Proof. By Corollary 8.12, statements i) and ii) above are equivalent to statements i) and ii) of Theorem 9.5, respectively. □

As a particular example of this result we have:

148

<u>Corollary 9.7.</u> Let T be an invertible operator and assume that there are invertible operators S_1 and S_2 satisfying $\|S_1^{-1}TS_1\| \leq R$, $\|S_2^{-1}T^{-1}S_2\| \leq r^{-1}$, with $r \leq R$. Then there is an invertible operator S such that $\|S^{-1}TS\| \leq R$ and $\|S^{-1}T^{-1}S\| \leq r^{-1}$.

<u>Proof.</u> When $r < R$, this is a direct application of Corollary 9.6, by letting $X = \{z: r \leq |z| \leq R\}$.

If $r = R$, then $\|(T/R)^n\| \leq \|S_1^{-1}\| \cdot \|S_1\|$, and $\|(T/R)^{-n}\| \leq \|S_2^{-1}\| \cdot \|S_2\|$. Hence, by Corollary 8.4, there exists an invertible operator S such that $S^{-1}(T/R)S$ is a unitary. Thus, $\|S^{-1}TS\| \leq R$ and $\|S^{-1}T^{-1}S\| \leq r^{-1}$.   □

<u>Theorem 9.8.</u> Let X be a nice n-holed set with canonical decomposition $\{X_i\}_{i=0}^{n}$ and (complete) decomposition constant K. If $T \in L(H)$ and each X_i is a spectral set for T, then X is a (complete) K-spectral set for T.

<u>Proof.</u> Since each X_i is a D-set, the hypothesis that X_i is a spectral set implies that T has a normal ∂X_i-dilation, by Theorem 4.4. That is, X_i is a complete spectral set for T. The remainder of the proof is a direct application of Propositions 9.3 and 9.4.   □

Let $X = \{z: r \leq |z| \leq R\}$, $r < R$, be an annulus, so that the canonical decomposition of X is $X_0 = \{z: |z| \leq R\}$ and $X_1 = \{z: r \leq |z|\}$. We have that X_0 is a spectral set for T if and only if $\|T\| \leq R$, and X_1 is a spectral set for T if and only if $\|T^{-1}\| \leq r^{-1}$.

We have seen in an earlier example that $\|T\| \leq R$ and $\|T^{-1}\| \leq r^{-1}$ is not enough, in general, to guarantee that the annulus is a spectral set for T. However, by the above theorem, the annulus will be a complete K-spectral set for T and so up to conjugation by a similarity T will have a normal ∂X-dilation.

It is not difficult to see from the proof of Proposition 9.4, that $2 + \dfrac{R + r}{R - r}$ gives an upper bound on the complete decomposition constant for the canonical decomposition of X. It is interesting to contrast this with [98, Proposition 23], where it is shown that the decomposition constant for this annulus is bounded by $2 + \left(\dfrac{R + r}{R - r}\right)^{\frac{1}{2}}$. However, that proof does not appear to generalize to matrices of analytic functions.

Precise values for the decomposition constant and complete decomposition, are not known, even for an annulus. Neither is it known whether these constants are achieved by operators. That is, for T an operator with $\|T\| \leq R$, $\|T^{-1}\| \leq r^{-1}$, let $\rho: R(X) \to L(H)$ be the homomorphism given by $\rho(f) = f(T)$, where $X = \{z: r \leq |z| \leq R\}$, and set

$$K = \sup \{\|\rho\|: \|T\| \leq R, \|T^{-1}\| \leq r^{-1}\},$$

and

$$K_c = \sup \{\|\rho\|_{cb}: \|T\| \leq R, \|T^{-1}\| \leq r^{-1}\}.$$

By Theorem 9.6, K and K_c will be less than or equal to the decomposition constant and complete decomposition for X, respectively, but it is not known if these inequalities are strict or are in fact equalities.

There are sets besides nice n-holed sets for which these decomposition techniques are valuable. For example, the sets

$$X_o = \{z: |z| \leq 1 \text{ and } |z - \tfrac{1}{2}| \geq \tfrac{1}{2}\}$$

and

150

$$X_1 = \{z: |z + \tfrac{1}{2}| \geq \tfrac{1}{4}\}$$

are D-sets [30, VI.11.11], and the proof of Proposition 9.4 can be suitably modified to show that these sets define a complete decomposition of $X = X_0 \cap X_1$. Consequently, the conclusions of Theorems 9.5 and 9.8 apply to these sets as well.

Another approach to finitely-connected regions involves the concept of a hypo-Dirichlet algebra. If X is a compact, Hausdorff space, then any subalgebra of R of $C(X)$, which separates points on X and has the property that the closure of $R + \overline{R}$ is of finite codimension n in $C(X)$, is called a hypo-Dirichlet algebra on X of codimension n .

Gamelin [42, Theorem IV.8.3] shows that if X is a compact subset of $\mathbb{C}$ whose complement has n bounded components, then $R(X)$ is a hypo-Dirichlet algebra on ∂X of codimension m , with $m \leq n$. In fact, if points $\{z_j\}_{j=1}^n$ are chosen, one from each bounded component of the complement of X , then the span of $R(X)$, $\overline{R(X)}$ and $\{\ln |z - z_j|\}_{j=1}^n$ is dense in $C(\partial X)$.

We shall show that unital, contractive homomorphisms of hypo-Dirichlet algebras are necessarily completely bounded.

Lemma 9.9. Let A be a unital C^*-algebra and let $S \subseteq A$ be an operator system of codimension n . Then for every $\varepsilon > 0$, there exists a positive map $\phi: A \to S$ and a positive linear functional s on S such that $\|\phi\| \leq n + 1 + \varepsilon$, $\|s\| \leq n + \varepsilon$, and $\phi(f) = f + s(f) \cdot 1$ for $f \in S$.

Proof. Let $\pi: A \to A/S$ be the quotient map. It is not difficult to show that there exist self-adjoint linear functionals, $\ell_1', \ell_2', \ldots, \ell_n'$, on A/S and self-adjoint elements, h_1', h_2', ..., h_n', in A/S which form a basis, such that $\|\ell_1'\| = \|h_i'\| = 1$ and $\ell_i(h_j') = \delta_{i,j}$, the Kronecker delta. Let h_1, h_2, ..., h_n be self-adjoint elements in A such that $\pi(h_i) = h_i'$ and $\|h_i\| \leq 1 + \varepsilon/n$. Also, let $\ell_i = \ell_i' \circ \pi$, so that $S = \{f \in A: \ell_i(f) = 0, \; i = 1, 2, \ldots, n\}$ with $\|\ell_i\| = 1$ and $\ell_i(h_j) = \delta_{i,j}$.

By Exercise 7.14, we can write $\ell_i = p_i - q_i$, where $\|p_i\| \leq 1$, $\|q_i\| \leq 1$, and $\|p_i + q_i\| \leq 1$, with p_i and q_i positive linear functionals in A. For every g in A,

$$f = g - \Sigma_{i=1}^n \ell_i(g)h_i$$

is in S. We define a positive map $\phi: A \to S$ by

$$\phi(g) = g + \Sigma_{i=1}^n q_i(g)(\|h_i\| + h_i) + \Sigma_{i=1}^n p_i(g)(\|h_i\| - h_i).$$

Since each of the three expressions defines a positive map, we need only check that the range of ϕ is contained in S. An easy calculation shows for $g = f + \Sigma_{i=1}^n \ell_i(g)h_i$, that $\phi(g) = f + \Sigma_{i=1}^n (p_i(g) + q_i(g))\|h_i\|$. If we set $s(g) = \Sigma_{i=1}^n (p_i(g) + q_i(g))\|h_i\|$, then

$$\|s\| \leq \Sigma_{i=1}^n \|p_i + q_i\| \|h_i\| \leq n + \varepsilon$$

and $\phi(f) = f + s(f) \cdot 1$ for f in S. Finally, since ϕ is positive, we have $\|\phi\| = \|\phi(1)\| = 1 + s(1) \leq n + 1 + \varepsilon$.

Theorem 9.10. If $A \subseteq C(X)$ is a hypo-Dirichlet algebra of codimension n, and $\rho: A \to L(H)$ is a unital contraction, then $\|\rho\|_{cb} \leq 2n + 1$.

Proof. Let $\varepsilon > 0$, let S be the closure of $A + \bar{A}$ in $C(X)$,

152

and let $\phi: C(X) \rightarrow S$ and s be as in the previous Lemma. If we extend
ρ to $\tilde{\rho}: S \rightarrow L(H)$ by $\tilde{\rho}(f + \bar{g}) = \rho(f) + \rho(g)^*$, then $\tilde{\rho}$ will be positive
by Proposition 2.12. Thus, $\tilde{\rho} \circ \phi: C(X) \rightarrow L(H)$ is positive and hence
completely positive. Finally, for f in A , we have

$$\rho(f) = \tilde{\rho} \circ \phi(f) - s(f) \cdot 1_H$$

so that

$$\|\rho\|_{cb} \leq \|\tilde{\rho} \circ \phi\|_{cb} + \|s\|_{cb} = \|\tilde{\rho} \circ \phi(1)\| + \|s(1)\| \leq 2n + 1 + 2\varepsilon ,$$

since $\tilde{\rho} \circ \phi$ and s are completely positive.

Corollary 9.11. If $A \subseteq C(X)$ is a hypo-Dirichlet algebra of codimension n , and $\rho: A \rightarrow L(H)$ is a unital contractive homomorphism, then ρ
is similar to a homomorphism that dilates to C(X) . Furthermore, the
similarity S may be chosen such that $\|S\| \cdot \|S^{-1}\| \leq 2n + 1$.

Corollary 9.12. Let X be a compact subset of $\mathbb{C}$ such that X is
a spectral set for T in L(H) . If R(X) is a hypo-Dirichlet algebra
of codimension n on ∂X , then there exists an invertible operator S
in L(H) with $\|S^{-1}\| \cdot \|S\| \leq 2n + 1$ such that $S^{-1}TS$ has a normal
∂X-dilation.

It is interesting to contrast Corollary 9.12 when X is a nice n-holed
set with Theorem 9.8. If X is a spectral set for T , then both theorems
allow us to deduce that X is a complete K-spectral set for T . Corollary
9.12 gives an upper bound on K of (2n + 1) , while the bound on K
in Theorem 9.8 comes from the complete decomposition constant for X

relative to the canonical decomposition of X . The proof of Proposition

9.4 gives a bound on the complete decomposition constant, but the bound

one obtains in this manner is always larger than (2n + 1) . However, it

should be recalled that the hypotheses of Theorem 9.8 are considerably

weaker than those of Corollary 9.12.

NOTES

Many of the results of this chapter were obtained in [36].

Exercise 9.3 is from [122].

EXERCISES

9.1 Let T be in $L(H)$ and let $X = \{z: Re\ (z) \geq 0\}$. Prove that the
following are equivalent:

i) X is a complete spectral set for T .

ii) X is a spectral set for T .

iii) $C = (T - 1)(T + 1)^{-1}$ is a contraction and $1 \notin \sigma(C)$.

iv) $Re\ (T) \geq 0$.

v) $(T - 1)^*(T - 1) \leq (T + 1)^*(T + 1)$.

Let X be a closed, proper subset of $\mathbb{C}$, $T \in L(H)$, and

suppose X is a spectral set for T . We say that T has a normal

∂X-dilation provided that there exists a Hilbert space K containing

H and a normal operator N in $L(K)$, with $\sigma(N) \subseteq \partial X$, such that

$f(T) = P_H f(N)|_H$.

154

9.2 Let $X = \{z: \text{Re } (z) \geq 0\}$ be a spectral set for T .

    i) Prove that T has a normal ∂X-dilation if and only if the minimal unitary dilation U of the contraction $C = (T - 1)(T + 1)^{-1}$ satisfies $1 \notin \sigma(U)$.

    ii) Give an example of an operator T such that X is a complete spectral set for T , but T has no normal ∂X-dilation.

    iii) Let $\rho: R(X) \to L(H)$ be given by $\rho(f) = f(T)$. Show that if X is a complete spectral set for T , then ρ has a $C(\partial \hat{X})$-dilation.

9.3 Let X be a proper, closed subset of $\mathbb{C}$, $\lambda \in \hat{X}$, and $T \in L(H)$ with $\sigma(T) \subseteq X$. Set $R_\lambda(X) = \{f \in R(X): f(\lambda) = 0\}$.

    i) (Williams) Show that if $\|f(T)\| \leq \|f\|$ for all $f \in R_\lambda(X)$, then X is a spectral set for T .

    ii) Show that if $\|f(T)\| \leq K\|f\|$ for all $f \in R_\lambda(X)$, then X' is a $(2K + 1)$-spectral set.

    iii) Give an example where $\|f(T)\| \leq K\|f\|$ for all $f \in R$, but X is not a K-spectral set.

    iv)* If $\|f(T)\| \leq \|f\|$ for all $f \in M_n(R_\lambda(X))$ and all n , then is X a complete spectral set for T ?

9.4 Let X_i , $i = 1, \ldots, n$, be disjoint, compact sets and let X be their union. Show that if X is a K-spectral set for T , then T is similar to $T_1 \oplus \ldots \oplus T_n$ such that X_i is K_i-spectral for T_i .

10 Tensor products and joint spectral sets

In this chapter we develop some of the theory of tensor products of C^*-algebras and operator systems, and apply this theory to multi-variable dilation theory. An n-tuple of operators $(T_1, \ldots, T_n)$ is said to doubly commute provided that $T_i T_j = T_j T_i$ and $T_i T_j^* = T_j^* T_i$ for all $i \neq j$. We shall see that for doubly commuting operators, a natural setting for generalizing the theory of spectral sets from a single variable theory to a multi-variable theory is the theory of tensor products.

Let A and B be unital C^*-algebras. Then their algebraic tensor product can be made into a *-algebra by setting $(a \otimes b)^* = a^* \otimes b^*$ and extending linearly. By a cross norm, we mean a norm $\| \cdot \|_\gamma$ on $A \otimes B$ with the property that $\|a \otimes b\|_\gamma = \|a\| \|b\|$ for $a \in A$ and $b \in B$. By a C^*-cross norm, we mean a cross norm on $A \otimes B$ which also satisfies the C^*-algebra axioms, $\|xy\|_\gamma \leq \|x\|_\gamma \|y\|_\gamma$ and $\|x^* x\|_\gamma = \|x\|_\gamma^2 = \|x^*\|_\gamma$ for $x, y \in A \otimes B$. The completion of $A \otimes B$ with respect to a C^*-cross norm is a C^*-algebra which we denote by $A \otimes_\gamma B$.

In general, there are many possible C^*-cross norms on $A \otimes B$, but there are two that we shall be interested in, the maximal and minimal C^*-cross norms.

In order to construct the minimal C^*-cross norm, we recall the theory of tensor products of Hilbert spaces. Suppose that H and K are Hilbert spaces. It is well-known that if we set $<h \otimes k, h' \otimes k'> = <h,h'>_H <k,k'>_K$ and extend linearly, then we obtain an inner product on $H \otimes K$. The completion of $H \otimes K$ with respect to this inner product is a Hilbert space which we still denote by $H \otimes K$. If T and S are operators on H and K, respectively, then setting $(T \otimes S)(h \otimes k) = (Th) \otimes (Sk)$ extends to a bounded, linear operator on $H \otimes K$ with $\|T \otimes S\| = \|T\| \, \|S\|$.

If $\pi_1: A_1 \to L(H_1)$ and $\pi_2: A_2 \to L(H_2)$ are unital *-homomorphisms, then we obtain a unital, *-preserving homomorphism

$\pi_1 \otimes \pi_2: A_1 \otimes A_2 \to L(H_1 \otimes H_2)$ by setting $\pi_1 \otimes \pi_2(a \otimes b) = \pi_1(a) \otimes \pi_2(b)$.
Thus, if for $x \in A_1 \otimes A_2$, we set

$$\|x\|_{min} = \sup \{\|\pi_1 \otimes \pi_2(x)\|: \pi_i: A_i \to L(H_i) \text{ unital, *-homomorphism}, \ i = 1,2\},$$

then we obtain a C^*-cross norm on $A_1 \otimes A_2$. This norm is called the minimal (or spatial) norm and the completion of $A_1 \otimes A_2$ in this norm is denoted $A_1 \otimes_{min} A_2$. Note that it has the property that if $\pi_i: A_i \to L(H_i)$, $i = 1, 2$, are any unital *-homomorphisms, then $\pi_1 \otimes \pi_2: A_1 \otimes A_2 \to L(H_1 \otimes H_2)$ can be extended, by continuity, to a unital *-homomorphism of $A_1 \otimes_{min} A_2$, denoted $\pi_1 \otimes_{min} \pi_2$.

The following result is quite deep, and explains the name of this C^*-cross norm. For a proof, see [116, Theorem IV.4.19].

Theorem 10.1 (Takesaki). Let A_1 and A_2 be unital C^*-algebras. If γ is a C^*-cross norm on $A_1 \otimes A_2$, then $\|x\|_{min} \leq \|x\|_\gamma$ for all $x \in A_1 \otimes A_2$.

Corollary 10.2. Let A_1 and A_2 be unital C^*-algebras. If $\pi_i: A_i \to L(H_i)$ are 1 - 1, unital *-homomorphisms, then for $x \in A_1 \otimes A_2$, $\|x\|_{min} = \|\pi_1 \otimes \pi_2(x)\|$.

Proof. By definition, $\|x\|_{min} \geq \|\pi_1 \otimes \pi_2(x)\|$. But setting

157

$\|x\|_\gamma := \|\pi_1 \otimes \pi_2(x)\|$ defines a C^*-cross norm, from which the other inequality follows. □

Suppose that B_i , $i = 1, 2$, are unital C^*-algebras and that $A_i \subseteq B_i$, $i = 1, 2$, are unital C^*-subalgebras. For $x \in A_1 \otimes A_2 \subseteq B_1 \otimes B_2$, we have two possible definitions of $\|x\|_{min}$, depending on whether we view it as an element of $A_1 \otimes A_2$ or of $B_1 \otimes B_2$. However, if we fix $\pi_i: B_i \to L(H_i)$, $i = 1, 2$, unital, $1 - 1$, *-homomorphisms, then since their restrictions are also unital, $1 - 1$, *-homomorphisms of A_i , $i = 1, 2$, by Corollary 10.2, $\|x\|_{min} = \|\pi_1 \otimes \pi_2(x)\|$, independent of which algebra we regard it as an element of. This observation can perhaps best be summarized by saying that the natural inclusion of $A_1 \otimes A_2$ into $B_1 \otimes B_2$ extends to a *-isomorphism of $A_1 \otimes_{min} A_2$ onto the norm closure of $A_1 \otimes A_2$ in $B_1 \otimes_{min} B_2$. Because of this last fact, the minimal C^*-cross norm is also called the <u>injective</u> C^*-cross norm.

We use this observation to define a min norm on tensor products of operator spaces. Suppose that A_i , $i = 1, 2$, are unital C^*-algebras and that $S_i \subseteq A_i$, $i = 1, 2$, is a subspace. We then define the min norm on $S_1 \otimes S_2$ to be the restriction of the min norm on $A_1 \otimes A_2$ and let $S_1 \otimes_{min} S_2$ denote its completion in this norm, i.e., its closure in $A_1 \otimes_{min} A_2$. Note that if S_1 and S_2 are operator systems, then $S_1 \otimes_{min} S_2$ is also an operator system.

<u>Theorem 10.3</u>. Let A_i and B_i be unital C^*-algebras, let $S_i \subseteq A_i$ be a subspace, and let $L_i: S_i \to B_i$ be completely bounded, $i = 1, 2$. Then the linear map, $L_1 \otimes L_2: S_1 \otimes S_2 \to B_1 \otimes B_2$, defined by $(L_1 \otimes L_2)(a_1 \otimes a_2) = L_1(a_1) \otimes L_2(a_2)$, extends to a completely bounded map,

158

$L_1 \otimes_{min} L_2 \colon S_1 \otimes_{min} S_2 \to B_1 \otimes_{min} B_2$, with $\|L_1 \otimes_{min} L_2\|_{cb} = \|L_1\|_{cb}\|L_2\|_{cb}$.
If S_1 and S_2 are operator systems and L_1 and L_2 are completely positive, then $L_1 \otimes_{min} L_2$ is completely positive.

Proof. Let $B_i \subseteq L(H_i)$, $i = 1, 2$. If we can show that
$L_1 \otimes L_2 \colon S_1 \otimes S_2 \to L(H_1 \otimes H_2)$ is completely bounded in the min norm
with $\|L_1 \otimes L_2\|_{cb} = \|L_1\|_{cb}\|L_2\|_{cb}$, then we will be done by the injectivity
of the min norm. By the extension theorem for completely bounded maps,
we may extend L_i to $\tilde{L}_i \colon A_i \to L(H_i)$ with $\|\tilde{L}_i\|_{cb} = \|L_i\|_{cb}$, $i = 1, 2$.
Now applying the generalized Stinespring representation, we obtain unital
*-homomorphisms, $\pi_i \colon A_i \to L(K_i)$, and bounded operators, $V_i \colon H_i \to K_i$,
$W_i \colon H_i \to K_i$, with $\|V_i\| \|W_i\| = \|L_i\|_{cb}$, such that

$$\tilde{L}_i(a_i) = V_i^* \pi_i(a) W_i , \quad a_i \in A_i , \quad i = 1, 2 .$$

Consider $V_1 \otimes V_2 \colon H_1 \otimes H_2 \to K_1 \otimes K_2$, $W_1 \otimes W_2 \colon H_1 \otimes H_2 \to K_1 \otimes K_2$,
and $\pi_1 \otimes_{min} \pi_2 \colon A_1 \otimes_{min} A_2 \to L(K_1 \otimes K_2)$. We have that
$(V_1 \otimes V_2)^*(\pi_1 \otimes \pi_2)(a_1 \otimes a_2)(w_1 \otimes w_2) = \tilde{L}_1(a_1) \otimes \tilde{L}_2(a_2)$. Thus,
$\|L_1 \otimes L_2\|_{cb} \leq \|V_1 \otimes V_2\| \|W_1 \otimes W_2\| = \|L_1\|_{cb}\|L_2\|_{cb}$ on $S_1 \otimes S_2$. We leave
it to the reader to verify that $\|L_1\|_{cb}\|L_2\|_{cb} \leq \|L_1 \otimes L_2\|_{cb}$.

If S_1 and S_2 are operator systems and L_1 and L_2 are completely
positive, then we argue as above using the extension theorem for completely
positive maps. In this case, we find that $V_i = W_i$, $i = 1,2$, from
which the result follows. □

Corollary 10.4. Let S_1 and S_2 be operator spaces, $x \in S_1 \otimes S_2$.
Then

$$\|x\|_{min} = \sup \{\|L_1 \otimes L_2(x)\| \colon L_i \colon S_i \to L(H_i) , \|L_i\|_{cb} \leq 1 , i = 1, 2\} .$$

159

Furthermore, if S_1 and S_2 are operator systems, then we may take L_i to be completely positive, $i = 1, 2$.

The above corollary shows that the min norm on tensor products of operator spaces is invariant under completely isometric isomorphisms of those operator spaces.

<u>10.5. $C(X;A)$</u>. Let X be a compact, Hausdorff space and let A be a unital C^*-algebra. We let $C(X;A)$ denote the continuous functions from X into A , equipped with the norm $\|F\| = \sup\{\|F(x)\|: x \in X\}$ for $F \in C(X;A)$. It is easy to check that if we define multiplication, addition, and the *-operation pointwise, then $C(X;A)$ is a C^*-algebra.

Now define a *-homomorphism from $C(X) \otimes A$ into $C(X;A)$ by

$$\Sigma_{i=1}^n f_i \otimes a_i \to F(x) = \Sigma_{i=1}^n f_i(x)a_i .$$

It is straightforward that setting $\|\Sigma_{i=1}^n f_i \otimes a_i\|_\gamma = \|F\|$ defines a C^*-cross norm on $C(X) \otimes A$. Furthermore, a standard partition of unity argument shows that the image of $C(X) \otimes A$ is dense in $C(X;A)$. Thus, the above mapping extends to a *-isomorphism of $C(X) \otimes_\gamma A$ with $C(X;A)$.

We can now show that γ is actually the minimal C^*-cross norm. To see this, note that for each fixed $x \in X$, the map $f \to f(x)$ extends to a contractive, linear map on $C(X)$. Thus by Exercise 10.1, the map $f \otimes a \to f(x)a$ extends to a contractive linear map from $C(X) \otimes_{min} A$ to A , and so $\|\Sigma_{i=1}^n f_i \otimes a_i\|_{min} \geq \|\Sigma_{i=1}^n f_i(x)a_i\|$. This shows that the min C^*-cross norm is greater than the γ C^*-cross norm, and consequently they must be equal.

We note that in the particular case where $A = C(Y)$ for a compact,

Hausdorff space Y , then the usual identification of a continuous function on $X \times Y$ with a continuous function from X into $C(Y)$ is a *-isomorphism of $C(X;C(Y))$ and $C(X \times Y)$. Thus, $C(X \times Y)$ is *-isomorphic to $C(X) \otimes_{min} C(Y)$.

If $S_i \subseteq C(X_i)$ is a subspace, $i = 1, 2$, then the min norm on $S_1 \otimes S_2$ is just the norm one obtains by viewing an element of $S_1 \otimes S_2$ as a function on $X_1 \times X_2$. Furthermore, if S_1 and S_2 are operator systems, then an element of $M_n(S_1 \otimes S_2)$ will be positive if and only if it is a positive function on $X_1 \times X_2$.

We now turn our attention to the max norm. Let A and B be unital C^*-algebras and let $\pi_1: A \to L(H)$, $\pi_2: B \to L(H)$ be unital *-homomorphisms such that $\pi_1(a)\pi_2(b) = \pi_2(b)\pi_1(a)$ for all $a \in A$ and $b \in B$. We may then define a unital *-homomorphism $\pi: A \otimes B \to L(H)$ via
$\pi(x) = \Sigma_{i=1}^n \pi_1(a_i)\pi_2(b_i)$, where $x = \Sigma_{i=1}^n a_i \otimes b_i$. Conversely, if we're given a unital *-homomorphism $\pi: A \otimes B \to L(H)$ and we define $\pi_1(a) = \pi(a \otimes 1)$, $\pi_2(b) = \pi(1 \otimes b)$, then we obtain a pair of unital *-homomorphisms of A and B , respectively, with commuting ranges.

We define

$$\|x\|_{max} = \sup \{\|\pi(x)\|: \pi: A \otimes B \to L(H) \text{ unital *-homomorphism}\} .$$

Thus, if $\pi_1: A \to L(H)$, $\pi_2: B \to L(H)$, are unital *-homomorphisms with commuting ranges, then we will have a unital *-homomorphism
$\pi_1 \otimes_{max} \pi_2: A \otimes_{max} B \to L(H)$ satisfying $\pi_1 \otimes_{max} \pi_2(a \otimes b) = \pi_1(a)\pi_2(b)$.

Proposition 10.6. Let A and B be unital C^*-algebras, let $x \in A \otimes B$, and let γ be a C^*-cross norm on $A \otimes B$, then $\|x\|_\gamma \leq \|x\|_{max}$.

161

Proof. By the Gelfand-Naimark-Segal theorem, there is a unital
*-homomorphism, $\pi: A \otimes_\gamma B \to L(H)$, such that $\|x\|_\gamma = \|\pi(x)\| \leq \|x\|_{max}$,
by definition.   □

If we have C^*-algebras, $A_i \subseteq B_i$,   i = 1, 2 , then the natural
inclusion of $A_1 \otimes A_2 \subseteq B_1 \otimes B_2 \subseteq B_1 \otimes_{max} B_2$ induces a C^*-cross norm γ
on $A_1 \otimes A_2$. The C^*-algebra $A_1 \otimes_\gamma A_2$ can be identified with the
closure of $A_1 \otimes A_2$ in $B_1 \otimes_{max} B_2$. Thus, by Proposition 10.6, we have
a *-homomorphism $A_1 \otimes_{max} A_2 \to B_1 \otimes_{max} B_2$ which, in general, can be norm
decreasing. For this reason, the max norm is often referred to as the
projective C^*-cross norm.

Unfortunately, the max norm on tensor products is not compatable with
completely bounded maps. Huruya [59] has given an example of a completely
bounded map $L: A_1 \to A_2$ and a C^*-algebra B such that the
L ⊗ id: $A_1 \otimes B \to A_2 \otimes B$, defined by L ⊗ id(a ⊗ b) = L(a) ⊗ b . does not
even extend to a bounded map from $A_1 \otimes_{max} B$ into $A_2 \otimes_{max} B$. However,
it is the case that the tensor product of completely positive maps yields
a completely positive map on the tensor product in the max norm. This
fact is a consequence of the following commutant lifting theorem of
Arveson [3].

For a set $S \subseteq L(H)$, let S' = {T ∈ L(H): TS = ST for all S ∈ S} .
Note that S' is always an algebra and that when S is self-adjoint,
S' is self-adjoint.

Theorem 10.7 (Arveson). Let H and K be Hilbert spaces, let
$B \subseteq L(K)$ be a C^*-algebra containing 1_K , and let V: H → K be a bounded
linear transformation with BVH norm dense in K . Then for every

$T \in (V^{*}BV)'$, there exists a unique $T_1 \in B'$ such that $VT = T_1V$.
Furthermore, the map $T \rightarrow T_1$ is a *-homomorphism of $(V^{*}BV)'$ onto
$B' \cap \{VV^{*}\}'$.

   <u>Proof.</u> Let A_1, ..., A_n be in B , and let h_1, ..., h_n be in H .
Note that if a T_1 with the desired properties exists, then

$$(*) \qquad T_1(\Sigma_{i=1}^{n} A_iVh_i) \;=\; \Sigma_{i=1}^{n} A_iVTh_i \; .$$

Since the vectors appearing in the left-hand side of the above formula are
dense in K , this shows that such a T_1 , provided that it exists, is
necessarily unique.

   Thus, we need only prove that the above formula yields a well-defined,
bounded operator. Note that if P and Q are commuting positive operators
and x is a vector, then $<PQx,x> = <PQ^{\frac{1}{2}}x, Q^{\frac{1}{2}}x> \leq \|P\|<Qx,x>$. We have that

$$\| \Sigma_{i=1}^{n} A_iVTh_i \|^{2} \;=\; \Sigma_{i,j=1}^{n} <T^{*}V^{*}A_i^{*}A_jVTh_j, \; h_i>$$

$$= \; \Sigma_{i,j=1}^{n} <T^{*}TV^{*}A_i^{*}A_jVh_j, \; h_i> \;=\; <PQx,x> \; ,$$

where $Q = (V^{*}A_i^{*}A_jV)_{i,j=1}^{n}$, $x = h_1 \oplus \ldots \oplus h_n$, and P is the diagonal
$n \times n$ operator, whose entries are $T^{*}T$. Thus, P and Q are positive
and commute, and so

$$\| \Sigma_{i=1}^{n} A_iVTh_i \|^{2} \;\leq\; \|P\|<Qx,x> \;=\; \|T\|^{2} \cdot \| \Sigma_{i=1}^{n} A_iVh_i \|^{2} \; .$$

This equation shows that the formula $(*)$ yields both a well-defined and
bounded operator with $\|T_1\| \leq \|T\|$.

   From formula $(*)$ it is clear that the map π , given by $\pi(T) = T_1$,
is a homomorphism into B' . To see that it is a *-homomorphism, calculate

163

$$\langle \pi(T)A_1Vh_1, A_2Vh_2 \rangle = \langle V^*A_2^*A_1VTh_1, h_2 \rangle$$

$$= \langle TV^*A_2^*A_1Vh_1, h_2 \rangle = \langle A_1Vh_1, A_2VT^*h_2 \rangle = \langle A_1Vh_1, \pi(T^*)A_2Vh_2 \rangle .$$

Since linear combinations of vectors of the above form are dense in K, we have that $\pi(T)^* = \pi(T^*)$.

To see that T_1 is also in the commutant of VV^*, observe

$$T_1VV^* = VTV^* = V(VT^*)^* = V(T_1^*V)^* = VV^*T_1^* .$$

Finally, to see that π is onto $B' \cap \{VV^*\}'$, let $X \in B' \cap \{VV^*\}'$ and let $V = PW$ be the polar decomposition of V, so that X commutes with $P = (VV^*)^{\frac{1}{2}}$. Let $T = W^*XW$, then for $A \in B$,

$$TV^*AV = W^*XWW^*PAPW = W^*XPAPW = W^*PAPXW = V^*APWW^*XW = V^*AVT ,$$

so that T is in the commutant of V^*BV. Also, $VT = PWW^*XW = PXW = XPW = XV$, so that $\pi(T) = X$. Thus, the *-homomorphism π is indeed onto $B' \cap \{VV^*\}'$. □

Note that T is in the kernel of the above *-homomorphism π if and only if $VT = 0$. Thus, if V has trivial kernel, then π is a *-isomorphism. It is also worthwhile to note that even when the map π has a kernel, the map $\theta(X) = W^*XW$ defines a completely positive splitting of π, i.e., $\pi \circ \theta(X) = X$.

Theorem 10.8. Let A_1, A_2, and B be unital C^*-algebras, and let $\theta_i: A_i \to B$ be completely positive maps, $i = 1, 2$, with commuting ranges. Then there exists a completely positive map $\theta_1 \otimes_{max} \theta_2: A_1 \otimes_{max} A_2 \to B$ with $\theta_1 \otimes_{max} \theta_2(a_1 \otimes a_2) = \theta_1(a_1)\theta_2(a_2)$.

<u>Proof</u>. Clearly, we may assume that $B = L(H)$. Let (π_1, V_1, K_1) be

a minimal Stinespring representation of θ_1 , and let

$\gamma_1 \colon (V_1^* \pi_1(A_1)V_1)' \to \pi_1(A_1)' \cap \{V_1 V_1^*\}'$ be the *-homomorphism of

Theorem 10.7. We then have that $\pi_1 \colon A_1 \to L(K_1)$, $\tilde{\theta}_2 = \gamma_1 \circ \theta_2 \colon A_2 \to L(K_1)$,

are completely positive maps with commuting ranges.

Let $V_1 = P_1 W_1$ be the polar decomposition of V_1 . By the remarks

following Theorem 10.7, $\theta_2(a_2) - W_1^* \tilde{\theta}_2(a_2)W_1$ is in the kernel of γ_1 and

consequently, $V_1(\theta_2(a_2) - W_1^* \tilde{\theta}_2(a_2)W_1) = 0$. Thus,

$V_1 \theta_2(a_2) = V_1 W_1^* \tilde{\theta}_2(a_2)W_1 = P_1 \tilde{\theta}_2(a_2)W_1 = \tilde{\theta}_2(a_2)P_1 W_1 = \tilde{\theta}_2(a_2)V_1$, since

$\tilde{\theta}_2(a_2)$ commutes with $V_1 V_1^* = P_1^2$ and hence with P_1 . Hence, we have

that $V_1^* \pi_1(a_1)\tilde{\theta}_2(a_2)V_1 = V_1^* \pi_1(a_1)V_1 \theta_2(a_2) = \theta_1(a_1)\theta_2(a_2)$.

Repeating the above argument, we let (π_2, K_2, V_2) be a minimal

Stinespring representation of $\tilde{\theta}_2$ and let

$\gamma_2 \colon (V_2^* \pi_2(A_2)V_2)' \to \pi_2(A_2)' \cap \{V_2 V_2^*\}'$. Then $\tilde{\pi}_1 = \gamma_2 \circ \pi_1 \colon A_1 \to L(K_2)$

is a *-homomorphism whose range commutes with $\pi_2(A_2)$. Also,

$V_2^* \pi_2(a_2)\tilde{\pi}_1(a_1)V_2 = \tilde{\theta}_2(a_2)\pi_1(a_1)$.

Finally, by the universal property of the max norm, we have a

*-homomorphism, $\pi \colon A_1 \otimes_{max} A_2 \to L(K_2)$ with $\pi(a_1 \otimes a_2) = \tilde{\pi}_1(a_1)\pi_2(a_2)$.

Let $V \colon H \to K_2$ be defined by $V = V_2 V_1$, so that $\theta \colon A_1 \otimes_{max} A_2 \to L(H)$,

defined by $\theta(x) = V^* \pi(x)V$, is completely positive. Finally,

$\theta(a_1 \otimes a_2) = V_1^* V_2^* \tilde{\pi}_1(a_1)\pi_2(a_2)V_2 V_1 = V_1^* \pi_1(a_1)\tilde{\theta}_2(a_2)V_1 = \theta_1(a_1)\theta_2(a_2)$. Thus,

$\theta_1 \otimes_{max} \theta_2 = \theta$ is the desired completely positive map. $\square$

Because of the different properties of the max and min norm, it is

important to know when they coincide, that is, when there is a unique

C^*-cross norm on $A \otimes B$ A C^*-algebra A which has the property that the

max and min C^*-cross norms coincide for every unital C^*-algebra B is called nuclear. There is a deep and elegant theory characterizing these C^*-algebras. See Lance [69] for an excellent survey. For our purposes it will be enough to know that commutative C^*-algebras are nuclear. It is also valuable to note that M_n is nuclear (Exercise 10.4), so that the norm we defined on $M_n(A)$ is the unique C^*-cross norm.

Proposition 10.9. Let X be a compact, Hausdorff space, then $C(X)$ is nuclear.

Proof. Let B be a unital C^*-algebra, and let $\pi_1: C(X) \to L(H)$, $\pi_2: B \to L(H)$, be *-homomorphisms with commuting ranges. It will be sufficient to fix $\sum_{i=1}^{n} f_i \otimes b_i$ in $C(X) \otimes B$ and show that its max and min norm coincide. We have shown earlier (10.5) that
$$\| \textstyle\sum_{i=1}^{n} f_i \otimes b_i \|_{min} = \sup \{ \| \textstyle\sum_{i=1}^{n} f_i(x) b_i \| : x \in X \} .$$
Let E be the $L(H)$-valued spectral measure associated with the *-homomorphism π_1. If B is a Borel set in X, then the projection $E(B)$ will commute with $\pi_2(B)$, since $\pi_2(B)$ commutes with $\pi_1(C(X))$. This implies that $\| \sum_{i=1}^{n} \pi_1(f_i)\pi_2(b_i) \| = \sup \{ \| E(B)(\sum_{i=1}^{n} \pi_1(f_i)\pi_2(b_i)) \| ,$ $\| E(X \backslash B)(\sum_{i=1}^{n} \pi_1(f_i)\pi_2(b_i)) \| \}$. Consequently, if $\{U_\lambda\}$ is any open cover of X,

$$\| \textstyle\sum_{i=1}^{n} \pi_1(f_i)\pi_2(b_i) \| = \sup_{\lambda} \{ \| E(U_\lambda)(\textstyle\sum_{i=1}^{n} \pi_1(f_i)\pi_2(b_i)) \| \} .$$

Now fix $\varepsilon > 0$, and for each $x \in X$, choose an open neighborhood U_x of x, such that $|f_i(x) - f_i(y)| < \varepsilon$, for $y \in U_x$, $i = 1, \ldots, n$. This implies that $\| E(U_x)\pi_1(f_i) - f_i(x)E(U_x) \| < \varepsilon$. Thus, $\| E(U_x)(\sum_{i=1}^{n} (\pi_1(f_i) - f_i(x))\pi_2(b_i)) \| < \varepsilon \, (\|b_i\| + \ldots + \|b_n\|)$. Since the collection $\{U_x\}$ forms an open cover, we have

$$\|\Sigma_{i=1}^{n} \ \pi_1(f_i)\pi_2(b_i)\| \leq \sup\{\|\Sigma_{i=1}^{n} \ f_i(x)\pi_2(b_i)\|: x \in X\} + \varepsilon(\|b_1\| + \ldots + \|b_n\|)$$

$$\leq \|\Sigma_{i=1}^{n} \ f_i \otimes b_i\|_{min} + \varepsilon(\|b_1\| + \ldots + \|b_n\|) \ .$$

Finally, using the facts that ε was arbitrary and that the max norm is the supremum over all such π_1 and π_2 , we have $\|\Sigma_{i=1}^{n} \ f_i \otimes b_i\|_{max} \leq \|\Sigma_{i=1}^{n} \ f_i \otimes b_i\|_{min}$, which completes the proof. $\square$

We are now in a position to discuss some of the applications of the tensor theory to operator theory. Recall that a set of operators $\{T_i\}$ is said to doubly commute if $T_i^*T_j = T_jT_i^*$ and $T_iT_j = T_jT_i$ for $i \neq j$. This is equivalent to requiring that the C^*-algebras generated by each of these operators commutes with the C^*-algebra generated by any of the other operators, but does not require that each of these C^*-algebras be commutative.

Theorem 10.10 (Sz.-Nagy-Foias). Let $\{T_i\}_{i=1}^{n}$ be a doubly commuting family of contractions on a Hilbert space H . Then there exists a Hilbert space K containing H as a subspace, and a doubly commuting family of unitary operators $\{U_i\}_{i=1}^{n}$ on K such that

$$T_1(k_1) \ \cdots \ T_n(k_n) \ = \ P_H U_1^{k_1} \cdots U_n^{k_n}|_H \ , \qquad \text{where} \qquad \begin{cases} T^k & , \ k > 0 \\ T^{*-k} & , \ k < 0 \ . \end{cases}$$

Moreover, if K is the smallest reducing subspace for the family $\{U_i\}_{i=1}^{n}$ containing H , then $\{U_i\}_{i=1}^{n}$ is unique up to unitary equivalence. That is, if $\{U_i'\}_{i=1}^{n}$ and K' are another such set, then there is a unitary $W: K \rightarrow K'$ leaving H fixed such that $WU_iW^* = U_i'$, $i = 1, \ldots, n$.

<u>Proof</u>. First, assume that $n = 2$. Then we have completely positive maps, $\theta_i: C(\Pi) \to L(H)$ defined by $\theta_i(p + \bar{q}) = p(T_i) + q(T_i)^*$, $i = 1, 2$. Since the range of θ_1 commutes with the range of θ_2, there is a completely positive map $\theta_1 \otimes_{max} \theta_2: C(\Pi) \otimes_{max} C(\Pi) \to L(H)$ satisfying

$$\theta_1 \otimes_{max} \theta_2(f_1 \otimes f_2) = \theta_1(f_1)\theta_2(f_2) .$$

However, we have that $C(\Pi) \otimes_{max} C(\Pi)$ is *-isomorphic to $C(\Pi \times \Pi)$. Thus, we have a completely positive map $\theta: C(\Pi^2) \to L(H)$ with

$$\theta(z_1^{k_1} z_2^{k_2}) = T_1(k_1)T_2(k_2) ,$$ where z_1 and z_2 are the coordinate functions on Π^2. The result now follows for $n = 1$ by considering the Stinespring representation of θ.

For $n > 2$, by using the associativity of the tensor product (Exercise 10.7), and arguing as above, one obtains a map $\theta: C(\Pi^n) \to L(H)$ with $\theta(z_1^{k_1} \cdots z_n^{k_n}) = T_1(k_1) \cdots T_n(k_n)$. □

The above result, for $n = 1$, is weaker than Ando's dilation result [2], since we need to assume that the operators doubly commute.

For operator systems $S_i \subseteq A_i$, $i = 1, 2$, we wish to define a max norm. Because of the projective properties of the max norm, it it <u>not</u> sufficient to just consider the norm induced by the inclusion $S_1 \otimes S_2 \subseteq A_1 \otimes_{max} A_2$. Instead we take Theorem 10.8 as our defining property. If $\theta_i: S_i \to L(H)$, $i = 1, 2$, are maps with commuting ranges, we always have a well-defined map $\theta_1 \otimes \theta_2: S_1 \otimes S_2 \to L(H)$. For $(x_{i,j}) \in M_n(S_1 \otimes S_2)$, we set

$$\|(x_{i,j})\|_{max} = \sup \{ \|(\theta_1 \otimes \theta_2(x_{i,j}))\| : \theta_\ell: S_\ell \to L(H) , \ell = 1, 2\} ,$$

where θ_1 and θ_2 are unital, completely positive maps. By considering the direct sum of sufficiently many of the maps $\theta_1 \otimes \theta_2$, we can obtain

168

a unital map $\gamma: S_1 \otimes S_2 \to L(H)$ with the property that

$\|(x_{i,j})\|_{max} = \|(\gamma(x_{i,j}))\|$ for all $(x_{i,j}) \in M_n(S_1 \otimes S_2)$ and all n .

We then define $S_1 \otimes_{max} S_2$ to be the operator system that is the closure of $\gamma(S_1 \otimes S_2)$.

Note that if S_1 and S_2 were actually C^*-algebras, then by Theorem 10.8, the norms on $M_n(S_1 \otimes S_2)$ as above would correspond to the original definition.

We summarize the properties of $S_1 \otimes_{max} S_2$ as follows:

Proposition 10.11. Let S_1 and S_2 be operator systems, B a C^*-algebra, and let $\theta_i: S_i \to B$, $i = 1, 2$, be completely positive maps with commuting ranges. Then:

i) There exists a completely positive map

$\theta_1 \otimes_{max} \theta_2: S_1 \otimes_{max} S_2 \to B$ with

$\theta_1 \otimes_{max} \theta_2(a_1 \otimes a_2) = \theta_1(a_1)\theta_2(a_2)$.

ii) An element $(x_{i,j})$ in $M_n(S_1 \otimes_{max} S_2)$ is positive if and only if $(\theta_1 \otimes_{max} \theta_2(x_{i,j}))$ is positive for all such pairs θ_1 , θ_2 , and C^*-algebras B .

Proof. Exercise 10.5. ⬚

As with C^*-algebras, it is important to know when the min and max norms coincide on operator systems. If A is a nuclear C^*-algebra, then for every operator system S , the min and max norms on $A \otimes S$ coincide (Exercise 10.6). Thus, we define an operator system S to be nuclear if for every operator system T , the min and max norms coincide on $S \otimes T$.

We now have all the necessary concepts to discuss joint spectral sets and joint dilations. Let $X_i \subseteq \mathbb{C}$ be compact, $i = 1, \ldots, n$. We set $X = X_1 \times \ldots \times X_n$ and define $\partial_d X = \partial X_1 \times \ldots \times \partial X_n$. We let $R_d(X)$ denote the subalgebra of $C(\partial_d X)$ spanned by functions of the form $r_1(z_1) \cdots r_n(z_n)$, $r_i \in R(X_i)$, and let $R(X)$ denote the subalgebra of $C(\partial_d X)$ consisting of the rational functions,

$R(X) = \{p(z_1, \ldots, z_n/q(z_1, \ldots, z_n): p,q$ are polynomials, $q \neq 0$ on $X\}$.

The algebra $R_d(X)$ is contained in $R(X)$, is algebraically isomorphic to the tensor product $R(X_1) \otimes \ldots \otimes R(X_n)$, and, in general, is not dense in $R(X)$. If $T_i \in L(H)$ with $\sigma(T_i) \subseteq X_i$ and the set $\{T_i\}$ commutes, then we have a well-defined homomorphism, $\rho: R_d(X) \to L(H)$, given by $\rho(r_1 \cdots r_n) = r_1(T_1) \cdots r_n(T_n)$. We call X a joint K-spectral set for $\{T_i\}_{i=1}^n$ provided that $\|\rho\| \leq K$, and a complete joint K-spectral set provided that $\|\rho\|_{cb} \leq K$.

If there exists a family of commuting normals $\{N_i\}_{i=1}^n$ on a Hilbert space K , containing H with $\sigma(N_i) \subseteq \partial X_i$, then we call $\{N_i\}_{i=1}^n$ a joint normal $\partial_d X$-dilation of $\{T_i\}_{i=1}^n$ provided that $r_1(T_1) \cdots r_n(T_n) = P_H r_1(N_1) \cdots r_n(T_n)|_H$, for all $r_i \in R(X_i)$. The following is immediate.

Proposition 10.12. Let $\{T_i\}_{i=1}^n$ be a family of commuting operators. Then $\{T_i\}_{i=1}^n$ has a joint normal $\partial_d X$-dilation if and only if X is a complete joint spectral set for $\{T_i\}_{i=1}^n$. There exists an invertible S with $\|S^{-1}\| \cdot \|S\| \leq K$, such that $\{S^{-1}T_iS\}_{i=1}^n$ has a joint normal $\partial_d X$-dilation if and only if X is a complete joint K-spectral set for $\{T_i\}_{i=1}^n$.

<u>Theorem 10.13.</u> Let $\{T_i\}_{i=1}^n$ be doubly commuting operators, and let X_i be a complete spectral set for T_i . If $R(X_i) + \overline{R(X_i)}$ is dense in $C(\partial X_i)$ for $i = 1, \ldots, n-1$, then $\{T_i\}_{i=1}^n$ has a joint normal $\partial_d X$-dilation.

<u>Proof.</u> Assume $n = 2$ and let S_i denote the closure of $R(X_i) + \overline{R(X_i)}$ in $C(\partial X_i)$ so that $S_1 = C(\partial X_1)$. By hypothesis, we have completely positive maps $\theta_i \colon S_i \to L(H)$ satisfying $\theta_i(f + \overline{g}) = f(T_i) + g(T_i)^*$, with commuting ranges. Thus, we have a completely positive map, $\theta_1 \otimes_{\max} \theta_2 \colon S_1 \otimes_{\max} S_2 \to L(H)$. But $S_1 \otimes_{\max} S_2 = S_1 \otimes_{\min} S_2$ and $S_1 \otimes_{\min} S_2$ is completely isometrically embedded in $C(\partial X_1) \otimes_{\min} C(\partial X_2) = C(\partial_d X)$.

Thus, for $(f_{i,j}) \in M_n(R_d(X))$,
$\|(\rho(f_{i,j}))\| = \|(\theta_1 \otimes_{\max} \theta_2(f_{i,j}))\| \leq \|(f_{i,j})\|_{\max} = \|(f_{i.j})\|_{\min}$, where $\rho \colon R_d(X) \to L(H)$ is the homomorphism defined by $\rho(r_1(z_1)r_2(z_2)) = r_1(T_1)r_2(T_2)$. But since $\|(f_{i,j})\|_{\min}$ is just the norm of $(f_{i,j})$ in $M_n(C(\partial_d X))$, we have that X is a complete, joint spectral set for $\{T_1, T_2\}$.

The proof for $n > 2$ is analogous. $\square$

<u>Lemma 10.14.</u> Let X be a compact Hausdorff space, $S \subseteq C(X)$ an operator system of codimension n , and let T be another operator system. Then for $(a_{i,j}) \in M_k(S \otimes T)$, $\|(a_{i,j})\|_{\max} \leq (2n + 1)\|(a_{i,j})\|_{\min}$.

<u>Proof.</u> By Lemma 9.9, for any $\varepsilon > 0$ there is a completely positive map $\phi \colon C(X) \to S$ with the property that for $f \in S$, $\phi(f) = f + s(f) \cdot 1$, where s is a positive linear functional with $\|s\| \leq n + \varepsilon$,

$\|\phi\| \le n + 1 + \varepsilon$. Let $a = \Sigma_{i=1}^{k} f_i \otimes b_i \in S \otimes T$, and note that we have a completely positive map $\phi \otimes id: C(X) \otimes_{max} T \to S \otimes_{max} T$ with $\|\phi \otimes id\| = \|\phi \otimes id(1 \otimes 1)\| \le n + 1 + \varepsilon$. Note that a can also be regarded as an element of $C(X) \otimes T$. But since $C(X)$ is nuclear and since the min norm is injective, the max and min norms of a in $C(X) \otimes T$ and the min norm of a in $S \otimes T$ all coincide. Thus, we have that $\|\phi \otimes id(a)\|_{max} \le (n + 1 + \varepsilon) \|a\|_{min}$. But $\phi \otimes id(a) = a + \Sigma_{i=1}^{k} s(f_i)1 \otimes b_i$, and so

$$\|a\|_{max} \le \|\phi \otimes id(a)\|_{max} + \|\Sigma s(f_i)b_i\| \le (n + 1 + \varepsilon)\|a\|_{min} + (n + \varepsilon)\|a\|_{min} .$$

The last inequality follows by noting that $s \otimes_{min} id: S \otimes_{min} T \to T$ has $\|s \otimes_{min} id\| = \|s\|$. Thus, $\|a\|_{max} \le (2n + 1)\|a\|_{min}$, since ε was arbitrary.

The proof for matrices is similar and uses the complete positivity of all the maps. $\square$

Theorem 10.15. Let $\{T_i\}_{i=1}^{n+1}$ be doubly commuting operators with X_i a complete spectral set for T_i . If $R(X_i)$ is a hypo-Dirichlet algebra of codimension k_i on ∂X_i , $i = 1, \ldots, n$, then there exists a similarity S with $\|S\| \cdot \|S^{-1}\| \le (2k_1 + 1) \cdots (2k_n + 1)$ such that $\{S^{-1}T_iS\}_{i=1}^{n+1}$ has a joint, normal $\partial_d X$-dilation.

Proof. First assume that $n = 1$. Let $S_i = R(X_i) + \overline{R(X_i)}$ and let $\phi_i: S_i \to L(H)$ be defined by $\phi_i(f + \overline{g}) = f(T_i) + g(T_i)^*$. Then ϕ_1 and ϕ_2 have commuting ranges, and so there is a completely positive map $\phi_1 \otimes_{max} \phi_2: S_1 \otimes_{max} S_2 \to L(H)$ and $\|\phi_1 \otimes_{max} \phi_2\|_{cb} = 1$. But by the above

172

lemma, the map $\theta_1 \otimes \theta_1: S_1 \otimes S_2 \to L(H)$ will extend to be completely bounded in the min norm, with $\|\theta_1 \otimes_{\min} \theta_2\| \leq (2k_1 + 1)$. Thus, X is a complete, joint $(2k_1 + 1)$-spectral set for $\{T_1, T_2\}$ from which the result follows.

To argue for an arbitrary n , note that by repeated applications of Proposition 10.11 and the above lemma, the identity map on $S_1 \otimes \ldots \otimes S_n$ extends to a completely bounded map from $S_1 \otimes_{\min} \cdots \otimes_{\min} S_n$ to $S_1 \otimes_{\max} \cdots \otimes_{\max} S_n$, with $\|id \otimes \ldots \otimes id\|_{cb} \leq (2k_1 + 1) \cdots (2k_n + 1)$. The proof is now completed as in the n = 1 case . □

Theorem 10.16. Let $\{T_i\}_{i=1}^{n}$ be doubly commuting operators with X_i a spectral set for T_i . If $R(X_i)$ is a hypo-Dirichlet algebra of codimension k_i on ∂X_i , then there exists a similarity S with $\|S\| \cdot \|S^{-1}\| \leq (2k_1 + 1) \cdots (2k_n + 1)$ such that $\{S^{-1}T_iS\}_{i=1}^{n}$ has a joint, normal $\partial_d X$-dilation.

Proof. Fix $\varepsilon > 0$ and let $\theta_i: C(\partial X_i) \to S_i$, where S_i is the closure of $R(X_i) + \overline{R(X_i)}$, and s_i be as in the lemma. Since X_i is spectral for T_i , we have positive maps $\theta_i: S_i \to L(H)$ with commuting ranges. Hence $\alpha_i = \theta_i \circ \phi_i: C(\partial X_i) \to L(H)$ are completely positive. Set $\beta_i = \theta_i \circ s_i$. Again using the nuclearity of $C(\partial X_i)$, we may form the min tensor of any combination of the maps α_i and β_i and obtain a completely positive map on $C(\partial_d X)$. But for f_i in S_i , $\theta_i(f_i) = \alpha_i(f_i) - \beta_i(f_i)$. Hence on $S_1 \otimes \ldots \otimes S_n$, $\theta_1 \otimes \ldots \otimes \theta_n = (\alpha_1 - \beta_1) \otimes \ldots \otimes (\alpha_n - \beta_n)$, which when the right hand side is expanded, expresses $\theta_1 \otimes \ldots \otimes \theta_n$ as a difference of sums of completely positive maps on $C(\partial_d X)$. Computing the sum of the norms of

173

each of these maps and letting ε tend to 0 yields the result. □

We close this chapter with one final result. Although the hypotheses of the theorem look quite restrictive, many operators that arise in operator theory have the property that the C^*-algebra that they generate is nuclear. For example, Toeplitz operators, subnormal operators, and essentially normal operators usually generate nuclear C^*-algebras.

Theorem 10.17. Let $\{T_i\}_{i=1}^{n+1}$ be doubly commuting operators with X_i completely K_i-spectral for T_i . If the C^*-algebra generated by each T_i is nuclear, $i = 1, \ldots, n$, then there exists a similarity S with $\|S\| \cdot \|S^{-1}\| \le K_1 \cdots K_{n+1}$ such that $\{S^{-1}T_iS\}_{i=1}^{n+1}$ has a joint, normal $\partial_d X$-dilation.

Proof. By hypothesis, the map $\rho_i : R(X_i) \to L(H)$ given by $\rho_i(f) = f(T_i)$ is completely bounded. Extend ρ_i to $\tilde{\theta}_i : C(\partial X_i) \to L(H)$ with $\|\tilde{\rho}_i\|_{cb} = \|\rho_i\|_{cb}$. We may then form

$\theta_1 \otimes_{min} \cdots \otimes_{min} \theta_n = \theta : C(\partial_d X) \to L(H) \otimes_{min} \cdots \otimes_{min} L(H)$ and $\|\theta\|_{cb} \le \|\rho_1\|_{cb} \cdots \|\rho_{n+1}\|_{cb}$. However, for $f \in R_d(X)$,

$\theta(f) = \rho_1 \otimes \cdots \otimes \rho_{n+1}(f)$ is in $C^*(T_1) \otimes \cdots \otimes C^*(T_{n+1})$, and the min norm on this latter algebra is the restriction of the min norm on $L(H) \otimes_{min} \cdots \otimes_{min} L(H)$. Hence we have that

$\rho = \rho_1 \otimes \cdots \otimes \rho_{n+1} : R_d(X) \to C^*(T_1) \otimes_{min} \cdots \otimes_{min} C^*(T_{n+1})$ is completely bounded with $\|\rho\|_{cb} \le \|\theta\|_{cb}$. Now the min and max norms will agree on the tensor product, and the tensor product in the max norm maps into the C^*-algebra generated by $T_1 , \ldots, T_{n+1}$. Thus, we have that the map $\tilde{\rho} : R_d(X) \to L(H)$ defined by $\tilde{\rho}(f_1(z_1) \cdots f_{n+1}(z_{n+1})) = f_1(T_1) \cdots f_{n+1}(T_{n+1})$

174

is completely bounded, from which the result follows. $\square$

NOTES

For a more thorough treatment of tensor products of operator systems, see Choi and Effros [23]. In particular, they obtain an abstract characterization of operator systems.

Lance's survey article [69] and Takesaki's text [116] are two excellent sources for further study of tensor products and nuclearity.

See Dash [32] for some other results on joint spectral sets.

EXERCISES

10.1 Let A_i , $i = 1, 2$, be unital C^*-algebras and let $f: A_1 \to \mathbb{C}$ be a bounded linear functional. Prove that there exists a completely bounded map $F: A_1 \otimes_{min} A_2 \to A_2$ with $\|F\|_{cb} = \|f\|$ such that $F(a_1 \otimes a_2) = f(a_1)a_2$. If f is positive, prove that F is completely positive.

10.2 Let A and B be unital C^*-algebras. Prove that if there exists a constant c such that $\|x\|_{max} \leq c\|x\|_{min}$ for all $x \in A \otimes B$, then $\|x\|_{max} = \|x\|_{min}$.

10.3 Let A and B be unital C^*-algebras. Verify the following containments,

$$\{\Sigma_{i,j}\, a_{i,j} \otimes b_{i,j} : (a_{i,j}) \in M_n(A)^+, (b_{i,j}) \in M_n(B)^+\}$$

$$\subseteq (A \otimes B) \cap (A \otimes_{max} B)^+ \subseteq (A \otimes B) \cap (A \otimes_{min} B)^+ .$$

10.4 Prove that every finite dimensional C^*-algebra is nuclear.

10.5 Let S_i , $i = 1, 2$, be operator systems and let B be a unital C^*-algebra.

    i) Prove that if $\theta_i: S_i \to B$, $i = 1, 2$, are unital, completely positive maps with commuting ranges, then there exists a unital, completely contractive map,

$$\theta_1 \otimes_{max} \theta_2: S_1 \otimes_{max} S_2 \to B \text{ with}$$
$$\theta_1 \otimes_{max} \theta_2(a \otimes b) = \theta_1(a)\theta_2(b) .$$

    ii) Deduce that $\theta_1 \otimes_{max} \theta_2$ is also completely positive.

    iii) Let $\theta_i: S_i \to L(H)$, $i = 1, 2$, be completely positive maps with commuting ranges and let $\theta_i(1) = P_i$. Prove that there exist unital, completely positive maps.

$\tilde{\theta}_i: S_i \to L(H)$, $i = 1, 2$, with commuting ranges such that $\theta_i = P_i^{\frac{1}{2}}\tilde{\theta}_i P_i^{\frac{1}{2}}$, $i = 1, 2$.

    iv) Prove Proposition 10.11.

10.6 Let A be a C^*-algebra. Prove that if $A \otimes_{max} B = A \otimes_{min} B$ for every C^*-algebra B , then $A \otimes_{max} S = A \otimes_{min} S$ for every operator system S .

10.7 Let S_i , $i = 1, 2, 3$, be operator systems and let γ denote either the min or max norm.

    i) Prove that $S_1 \otimes_\gamma (S_2 \otimes_\gamma S_3)$ and $(S_1 \otimes_\gamma S_2) \otimes_\gamma S_3$ are completely isometrically isomorphic.

    ii) Prove that if S_1 and S_2 are nuclear, then $S_1 \otimes_{max} S_2 = S_1 \otimes_{min} S_2$ is nuclear.

10.8 (Holbrook-Sz.-Nagy) Let S and T be operators that doubly commute.

    i) Prove that if $S \in C_\rho$ and $T \in C_\sigma$, then $ST \in C_{\rho\sigma}$.

    ii) Prove that $w(ST) \leq \|S\| w(T)$.

    iii) Prove that $w(ST) \leq 2w(S)w(T)$, and give an example to show that this inequality is sharp.

Bibliography

1. J. Agler, Rational dilation on an annulus, Ann. of Math. 121 (1985), 537-564.

2. T. Ando, On a pair of commutative contractions, Acta Sci. Math. 24 (1963), 88-90.

3. W. B. Arveson, Subalgebras of C^*-algebras, Acta Math. 123 (1969), 141-224.

4. W. B. Arveson, Subalgebras of C^*-algebras II, Acta Math. 128 (1972), 271-308.

5. W. B. Arveson, An Introduction to C^*-algebras, Springer-Verlag, New York, 1976.

6. W. B. Arveson, Notes on extensions of C^*-algebras, Duke Math. J. 44 (1977), 329-355.

7. B. A. Barnes, The similarity problem for representations of a B^*-algebra, Mich. Math. J. 22 (1975), 25-32.

8. B. A. Barnes, When is a representation of a Banach algebra Naimark-related to a *-representation?, Pacific J. Math. 72 (1977), 5-25.

9. C. A. Berger, A strange dilation theorem, Notices Amer. Math. Soc. 12 (1965), 590. Abstract 625-152.

10. C. A. Berger and J. G. Stampfli, Norm relations and skew dilations, Acta Sci. Math., 28 (1967), 191-195.

11. C. A. Berger and J. G. Stampfli, Mapping theorems for the numerical range, Amer. J. Math., 89 (1967), 1047-1055.

12. A. Brown and C. Pearcy, Introduction to Operator Theory I: Elements of Functional Analysis, Springer-Verlag, New York, 1977.

13. J. W. Bunce, Representations of strongly amenable C^*-algebras, Proc. Amer. Math. Soc. 32 (1972), 241-246.

14. J. W. Bunce, The similarity problem for representations of C^*-algebras, Proc. Amer. Math. Soc. 81 (1981), 409-413.

15. J. W. Bunce, Approximating maps and a Stone-Weierstrass theorem for C^*-algebras, Proc. Amer. Math. Soc. 79 (1980), 559-563.

16. M. D. Choi, Positive linear maps on C^*-algebras, Canad. J. Math. 24 (1972), 520-529.

17. M. D. Choi, A Schwarz inequality for positive linear maps on C^*-algebras, Illinois J. Math. 18 (1974), 565-574.

18. M. D. Choi, Completely positive linear maps on complex matrices, Lin. Alg. and Appl., 10 (1975), 285-290.

19. M. D. Choi, Some assorted inequalities for positive linear maps on C^*-algebras, J. Operator Theory 4 (1980), 271-285.

20. M. D. Choi, Positive linear maps, in operator algebras and applications (R. V. Kadison, ed.), Proceedings of Symposia in Pure Mathematics, vol. 38, AMS, Providence, 1982.

21. M. D. Choi and E. G. Effros, The completely positive lifting problem for C^*-algebras, Ann. of Math. 104 (1976), 585-609.

22. M. D. Choi and E. G. Effros, Separable nuclear C^*-algebras and injectivity, Duke Math J. 43 (1976), 309-322.

23. M. D. Choi and E. G. Effros, Injectivity and operator spaces, J. Functional Analysis 24 (1977), 156-209.

24. M. D. Choi and E. G. Effros, Nuclear C^*-algebras and injectivity. The general case, Ind. Univ. Math. J. 26 (1977), 443-446.

25. M. D. Choi and E. G. Effros, Nuclear C^*-algebras and the approximation property, Amer. J. Math. 100 (1978), 61-79.

26. E. Christensen, Extensions of derivations, J. Funct. Anal. 27 (1978), 234-247.

27. E. Christensen, Extensions of derivations II, Math. Scand. 50 (1982), 111-122.

28. E. Christensen, On non-selfadjoint representations of operator algebras, Amer. J. Math. 103 (1981), 817-834.

29. A. Connes, Classification of injective factors, Annals of Math. 104 (1976), 585-609.

30. J. B. Conway, Subnormal Operators, Pitman, Boston, 1981.

31. M. J. Crabb and A. M. Davie, von Neumann's inequality for Hilbert space operators, Bull. London Math. Soc. 7 (1975), 49-50.

32. A. T. Dash, Joint spectral sets, Rev. Roumaine Math. Pures Appl. 16 (1971), 13-26.

33. J. Dixmier, Les moyennes invariante dans les semi-groupes et leurs applications, Acta Sci. Math. Szeged 12 (1950), 213-227.

34. J. Dixmier, C*-algebras, North-Holland, New York, 1977.

35. R. G. Douglas, Banach algebra techniques in operator theory, Academic Press, New York, 1972.

36. R. G. Douglas and V. I. Paulsen, Completely bounded maps and hypo-Dirichlet algebras, preprint.

37. N. Dunford and J. T. Schwartz, Linear operators I: General theory, Interscience, New York, 1958.

38. L. Fejer, Über trigonometrishe Polynome, Journal für Math 146 (1915), 53-82.

39. P. A. Fillmore, Notes on operator theory, Van Nostrand-Reinhold, New York, 1970.

40. S. R. Foguel, A counterexample to a problem of Sz.-Nagy, Pwc. Amer. Math. Soc. 15 (1964), 788-790.

41. C. Foias, Sur certains théorèmes de J. von Neumann, concernant les ensembles spectreaux, Acta Sci. Math. 18 (1957), 15-20.

42. T. Gamelin, Uniform algebras, Prentice-Hall, Englewood Cliffs, New Jersey, 1969.

43. D. Gaspar and A. Racz, An extension of a theorem of T. Ando, Michigan Math. J. 16 (1969), 377-380.

44. U. Haagerup, Solution of the similarity problem for cyclic representations of C*-algebras, Ann. of Math. 118 (1983), 215-240.

45. U. Haagerup, Decomposition of completely bounded maps on operator algebras, preprint.

46. U. Haagerup, Injectivity and decomposition of completely bounded maps, preprint.

47. D. W. Hadwin, Dilations and Hahn decompositions for linear maps, Canad. J. Math. 33 (1981), 826-839.

48. P. R. Halmos, On Foguel's answer to Nagy's question, Proc. Amer. Math. Soc. 15 (1964), 791-793.

49. P. R. Halmos, Ten problems in Hilbert space, Bull. Amer. Math. Soc. 76 (1970), 887-933.

50. E. Heinz, Ein v. Neumannscher Satz über beschränkte Operatoren im Hilbertschen Raum, Göttinger Nachr., 1952, 5-6.

51. D. A. Herrero, A Rota universal model for operators with multiply connected spectrum, Rev. Roum. Math. Pures et Appl. 21 (1976), 15-23.

52. D. A. Herrero, Approximation of Hilbert space operators, Pitman, Boston, 1982.

53. J. A. R. Holbrook, On the power bounded operators of Sz.-Nagy and Foias, Acta Sci. Math. 29 (1968), 299-310.

54. J. A. R. Holbrook, Multiplicative properties of the numerical radius in operator theory, J. Reine Angew. Math 237 (1969), 166-174.

55. J. A. R. Holbrook, Inequalities governing the operator radii associated with unitary ρ-dilations, Michigan Math. J. 18 (1971), 149-159.

56. J. A. R. Holbrook, Spectral dilations and polynomially bounded operators, Indiana Univ. Math. J. 20 (1971), 1027-1034.

57. J. A. R. Holbrook, Distortion coefficients for crypto-contractions, Lin. Alg. and Appl. 18 (1977), 229-256.

58. J. A. R. Holbrook, Distortion coefficients for crypto-unitary operators, Lin. Alg. and Appl. 19 (1978), 189-205.

59. T. Huruya, On compact completely bounded maps of C^*-algebras, Michigan Math. J. 30 (1983), 213-220.

60. T. Huruya, Linear maps between certain non-separable C^*-algebras, preprint.

61. T. Huruya, Decompositions of linear maps into non-separable C^*-algebras, preprint.

62. T. Huruya and J. Tomiyama, Completely bounded maps of C^*-algebras, J. Operator Theory 10 (1983), 141-152.

63. R. V. Kadison, On the orthogonalization of operator representations, Amer. J. Math. 77 (1955), 600-620.

64. R. V. Kadison, A generalized Schwarz inequality and algebraic invariants for C^*-algebras, Ann. Math. 56 (1952), 494-503.

65. T. Kato, Some mapping theorems for the numerical range, Proc. Japan Acad. 41 (1965), 652-655.

66. E. Kirchberg, C^*-nuclearity implies CPAP, Math. Nachr. 76 (1977), 203-212.

67. R. A. Kunze and E. M. Stein, Uniformly bounded representations and harmonic analysis of the 2 × 2 real unimodular group, Amer J. Math. 82 (1960), 1-62.

68. C. Lance, On nuclear C^*-algebras, J. Funct. Anal. 12 (1973), 157-176.

69. C. Lance, Tensor products and nuclear C^*-algebras, in operator algebras and applications (R. V. Kadison, ed), Proceedings of Symposia in

Pure Mathematics, vol. 38, AMS, Providence, 1982.

70. A. Lebow, On von Neumann's theory of spectral sets, J. Math. Anal. and Appl. 7 (1963), 64-90.

71. R. I. Loebl, Contractive linear maps on C^*-algebras. Michigan Math. J. 22 (1975), 361-366.

72. R. I. Loebl, A Hahn decomposition for linear maps, Pacific J. Math. 65 (1976), 119-133.

73. M. J. McAsey and P. S. Muhly, Representations of non-self-adjoint crossed products, preprint.

74. W. Mlak, Unitary dilations of contraction operators, Rozprawy Mat. 46 (1965), 1-88.

75. W. Mlak, Unitary dilations in case of ordered groups, Ann. Polon. Math. 17 (1966), 321-328.

76. W. Mlak, Positive definite contraction valued functions, Bull. Acad. Polon. Sci. Ser. Sci. Math. Astronom. Phys. 15 (1967), 509-512.

77. W. Mlak, Absolutely continuous operator valued representations of function algebras, Bull. Acad. Polon. Sci. Ser. Sci. Math. Astronom. Phys. 17 (1969), 547-550.

78. W. Mlak, Decompositions and extensions of operator valued representations of function algebras, Acta. Sci. Math. (Szeged) 30 (1969), 181-193.

79. W. Mlak, Decompositions of operator-valued representations of function algebras, Studia Math 36 (1970), 111-123.

80. M. A. Naimark, On a representation of additive operator set functions, C. R. (Doklady) Acad. Sci. URSS 41 (1943), 359-361.

81. M. A. Naimark, Positive definite operator functions on a commutative group, Bulletin (Izvestiya) Acad. Sci. URSS (ser. math.), 7 (1943), 237-244.

82. J. von Neumann, Eine spektraltheorie für allgemeine Operatoren eines unitären Raumes, Math. Nachr. 4 (1951), 258-281.

83. S. K. Parrott, Unitary dilations for commuting contractions, Pacific J. Math., 34 (1970), 481-490.

84. V. I. Paulsen, Completely bounded maps on C^*-algebras and invariant operator ranges, Proc. Amer. Math. Soc. 86 (1982), 91-96.

85. V. I. Paulsen, Every completely polynomially bounded operator is similar to a contraction, J. Funct. Anal. 55 (1984), 1-17.

86. V. I. Paulsen, Completely bounded homomorphisms of operator algebras, Proc. Amer. Math. Soc. 92 (1984), 225-228.

87. V. I. Paulsen and C.-Y. Suen, Commutant representations of completely bounded maps, J. Operator Theory 13 (1985), 87-101.

88. C. Pearcy, An elementary proof of the power inequality for the numerical radius, Mich. Math. J. 13 (1966), 289-291.

89. G. K. Pedersen, C^*-algebras and their automorphism groups, Academic Press, London, 1979.

90. F. Riesz and B. Sz.-Nagy, Functional Analysis, 2nd Edition, New York, 1955.

91. J. R. Ringrose, Automatic continuity of derivations of operator algebras, J. London Math. Soc. 5 (1972), 432-438.

92. G. C. Rota, On models for linear operators, Comm. Pure Appl. Math. 13 (1960), 468-472.

93. B. Russo and H. A. Dye. A note on unitary operators in C^*-algebras, Duke Math. J. 33 (1966), 413-416.

94. S. Sakai, C^*-algebras and W^*-algebras. Springer-Verlag, New York, 1971.

95. D. Sarason, On spectral sets having connected complement, Acta Sci. Math. 26 (1965), 289-299.

96. D. Sarason, Generalized interpolation in H^∞, Trans. Amer. Math. Soc. 127 (1969), 179-203.

97. J. J. Schaffer, On unitary dilations of contractions, Proc. Amer. Math. Soc. 6 (1955), p.322.

98. A. Shields, Weighted shift-operators and analytic function theory, in Topics in operator theory (C. Pearcy, ed.), American Mathematical Society, Providence, 1974.

99. R. R. Smith, Completely bounded maps between C^*-algebras, J. London Math. Soc. 27 (1983), 157-166.

100. R. R. Smith, Private communication.

101. R. R. Smith and J. D. Ward, Matrix ranges for Hilbert space operators, Amer. J. Math. 102 (1980), 1041-1081.

102. R. R. Smith and J. D. Ward, Locally isometric liftings from quotient C^*-algebras, Duke Math. J. 47 (1980). 621-631.

103. R. R. Smith and D. Williams, The decomposition property for C^*-algebras, preprint.

104. W. F. Stinespring, Positive functions on C^*-algebras, Proc. Amer. Math. Soc. 6 (1955), 211-216.

105. E. Stormer, Positive linear maps of C^*-algebras, Lecture Notes in Physics, vol. 29, pp. 85-106, Springer-Verlag, Berlin 1974.

106. E. Stormer, Extension of positive maps into B(H), preprint.

107. I. Suciu, Unitary dilations in case of partially ordered groups, Bull. Acad. Polon. Sci. (ser. math., astr. phys.), 15 (1967), 271-275.

108. C.-Y. Suen, Completely bounded maps on C^*-algebras, Proc. Amer. Math. Soc. 93 (1985), 81-87.

109. C.-Y. Suen, The unique representation of a self-adjoint bounded linear functional, preprint.

110. B. Sz.-Nagy, On uniformly bounded linear transformations in Hilbert space, Acta Sci. Math. Szeged 11 (1947), 152-157.

111. B. Sz.-Nagy, Sur les contractions de e´espace de Hilbert, Acta Sci. Math. 15 (1953), 87-92.

112. B. Sz.-Nagy, Completely continuous operators with uniformly bounded iterates, Magyar Tud. Akad. Mat. Kutató Int. Köze 4 (1959), 89-93.

113. B. Sz.-Nagy, Products of operators of class C_ρ, Rev. Roumaine Math. Pures Appl. 13 (1968), 897-899.

114. B. Sz.-Nagy and C. Foias, Harmonic analysis of operators on Hilbert Space, American Elsevier, New York, 1970.

115. T. Takasaki and J. Tomiyama, On the geometry of positive maps in matrix algebras, Math. Z. 184 (1983), 101-108.

116. M. Takesaki, Theory of operator algebras I, Springer-Verlag, Berlin, 1979.

117. J. Tomiyama, On the transpose map of matrix algebras, Proc. Amer. Math. Soc. 88 (1983), 635-638.

118. J. Tomiyama, Recent development of the theory of completely bounded maps between C^*-algebras, preprint.

119. N. Th. Varopoulos, On an inequality of von Neumann and an application of the metric theory of tensor products to operators theory, J. Funct. Anal. 16 (1974), 83-100.

120. D. Voiculescu, Norm-limits of algebraic operators, Rev. Roum. Math. Pures et Appl. 19 (1974), 371-378.

121. J. L. Walsh, The approximation of harmonic functions by harmonic polynomials and by harmonic rational functions, Bull. Amer. Math.

Soc. 35 (1929), 499-544.

122. J. P. Williams, Schwarz norms for operators, Pacific J. Math. 24 (1968), 181-188.

123. G. Wittstock, Ein operatorwertiger Hahn-Banach Satz, J. Funct. Anal. 40 (1981), 127-150.

124. G. Wittstock, On matrix order and convexity, Functional Analysis: Surveys and Recent Results, Math. Studies 90, p.p. 175-188, North-Holland, Amsterdam, 1984.

125. S. L. Woronowicz, Nonextendible positive maps, Comm. Math. Phys. 51 (1976), 243-282.

126. S. L. Woronowicz, Positive maps of low dimensional matrix algebras, Rep. Math. Phys. 10 (1976), 165-183.

Index

amenable group, 126
Ando, 21, 58, 59, 92
Arveson, 9, 38, 58, 81, 82, 162

B-dilation, 83
  minimal, 84
Berger, 38, 59, 62
  Foias and Lebow, 49
  Kato and Stampfli, 37
  and Stampfli, 38
Bimodule map, 60
Bunce, 59
BW-topology, 79

Canonical
  decomposition, 146
  shuffle, 97
C^*-cross norm, 156
  injective, 158
  maximal, 156
  minimal, 156, 157
  projective, 162
  spatial, 157
Choi, 20, 34, 38, 39, 40, 53, 58
  and Effros, 38, 93, 175
Christensen, 132, 138
Complete
  joint K-spectral set, 170
  K-spectral set, 133, 144
  spectral set, 84
Completely
  bounded, 5, 25
  contractive, 25
  isometric, 25
  polynomially bounded, 133
  positive, 5, 25
  positive definite, 53
Conditional expectation, 94
Connes, 93, 116
Crabbe and Davies, 21, 59, 92
Cross-norm, 93, 156
  injective, 93

Dash, 175
Decomposition constant, 146

Derivation, 130
  inner, 130
Dirichlet
  algebra, 49
  set, 146
Dixmier, 127, 137
Doubly commute, 156
D-set, 146

Fejer-Riesz, 11
Foguel, 138
Foias, 21

Gaspar and Racz, 92
Group C^*-algebra, 56

Haagerup, 116, 119, 125, 137, 138
Hadwin, 59, 137, 138
Halmos, 133, 138, 139
Heinz, 21
Herrero, 135
Holbrook, 134, 138, 177
Huruya, 117
Hypo-Dirichlet algebra, 72, 151

Injective
  C^*-algebra, 82
  C^*-cross norm, 158
  cross-norm, 93
  tensor product, 93
Invariant mean, 126

Joint
  K-spectral set, 170
  normal ∂_dX-dilation, 170
Jorgensen, 23

Kadison, 126, 137
Korovkin, 59, 61
Krein, 23
K-spectral set, 133, 144
Kunze and Stein, 137

Lance, 167, 175
Loebl, 108